Hawk Ridge

Hawk Ridge

Minnesota's Birds of Prey

Laura Erickson • *Illustrations by* Betsy Bowen

UNIVERSITY OF MINNESOTA PRESS

MINNEAPOLIS • LONDON

THE UNIVERSITY OF MINNESOTA PRESS gratefully acknowledges the generous assistance provided for the publication of this book by the Margaret W. Harmon Fund.

Published by the University of Minnesota Press
111 Third Avenue South, Suite 290
Minneapolis, MN 55401-2520
http://www.upress.umn.edu

A Cataloging-in-Publication record is available for this book from the Library of Congress.
ISBN 978-0-8166-8119-8

Book design by Brian Donahue / bedesign, inc.

Printed in Canada on acid-free paper

The University of Minnesota is an equal-opportunity educator and employer.

20 19 18 17 16 15 14 13 12   10 9 8 7 6 5 4 3 2 1

To Janet Green, who has given Hawk Ridge
so much of its past, present, and future L. E.

To my friends at the Tarpon Ice House,
where these paintings were first exhibited B. B.

Contents

Artist's Note

The hawk whistling overhead startles me . . . catches me up in the wonder of the freedom and fierceness and tender parenting of these raptors.

The color paintings in this book, acrylic on paper, were completed during the winter at an artists' compound on the Gulf Coast of the United States. Not far away I was able to observe several hawks and vultures up close in a rescue hospital at the University of Texas Marine Science Institute in Port Aransas, Texas. Ink drawings were created in the spring at Yonder Farm, just up the shore of Lake Superior from Hawk Ridge.

← ink cartridge

↖ Last drop!

Introduction

Every autumn, close to twenty thousand people gravitate to a stretch of unimproved road on a hill overlooking Lake Superior in Duluth, Minnesota, to watch thousands of individuals of twenty species of raptors pass by Hawk Ridge. Hawk migration is as certain as sunset, albeit seasonal rather than daily and with far less predictable timing. Heavy rains can obscure a sunset; they shut down hawk migration altogether. Both phenomena can overpower us with their compelling beauty, yet both meet the very definition of commonplace—they take place silently and regularly, with no impact on human events. Even the most avid nature lovers may not notice a spectacular sunset or an astounding flight of hawks unless they happen to be looking skyward at the right moment.

The sun was setting every day long before humans or hawks were on the scene to behold its splendor, but people may well have witnessed the very first hawk movements along Lake Superior. The ancient Laurel people, credited with creating many of the pictographs on rock faces along the Canadian North Shore and in the Boundary Waters Canoe Area Wilderness, began settling the Lake Superior region as the glaciers of the last ice age began to retreat, their meltwaters forming the lake as tundra and forest habitat slowly took form in the rubble. Breeding birds settled into developing habitats, and numbers of migrants grew apace. As they withdrew south in winter, the lake posed a huge barrier to migration. They couldn't go straight, so many birds turned right, or southwest, to make progress toward the south while avoiding the downdrafts and other treacherous conditions over water. This coincidentally allowed them to take advantage of the thermal air currents that form along shorelines and hold soaring birds aloft so effortlessly. It's likely that the first people to take notice of the annual streams of hawks passing over felt the same surge of joy and reassurance articulated by Rachel Carson when she wrote, "There is symbolic as well as actual beauty in the migration of birds. . . . There is something infinitely healing in the repeated refrains of nature—the assurance that dawn comes after night, and spring after the winter."

This book invites the reader to share in the time-honored, rhythmic annual tradition of hawk watching at Hawk Ridge. Twenty species of hawks have been seen at the ridge, including vultures, ospreys, eagles, harriers, kites, accipiters, buteos, and falcons. This book discusses each species' quirks of timing, daily as well as seasonal, points out challenges with identification, and provides life history information. The species are organized in the order of the American Ornithologists' Union's Checklist of North American Birds as updated through early 2012.

A History of Hawk Watching at Hawk Ridge

There is no more intrinsic economic value in hawk migration than in a sunset. It's now illegal to shoot hawks in the United States, but even before enactment of the Migratory Bird Act and its amendments, few people considered hawk meat a delicacy. Native Americans have long used raptor feathers for decorative and ceremonial purposes, but before guns were available, raptors would have been far more easily hunted on their seasonal territories than while migrating far overhead. After European settlement, for a time some local, state, and federal governments paid a bounty for the killing of hawks, and even without a monetary incentive, shooting hawks was long a popular pastime with Duluth residents who lacked understanding of nature's

complexities. "Hawk hill" was originally named by local gunners, who shot as many hawks as they could there. Members of the Duluth Bird Club lobbied the City of Duluth to outlaw the discharge of firearms within city limits. The resulting ordinance greatly decreased the slaughter.

When in 1951 local ornithologist Jack Hofslund and the Duluth Bird Club organized counts of the raptors passing over the ridge, ornithologists throughout the nation were astounded by the magnitude of the migration, the size of which had never been documented away from the coasts. Not even the world-famous Hawk Mountain in Pennsylvania approached Hawk Ridge's numbers of raptors, especially Broad-winged Hawks and Northern Goshawks.

In 1972, national attention as well as growing state and local interest led the newly incorporated Duluth Audubon Society to ask the Minnesota chapter of The Nature Conservancy for a loan to purchase and protect Hawk Ridge. The money from the loan was donated to the City of Duluth to purchase, under the city's Parks and Recreation Department, 135 acres of tax-forfeited land at the highest part of the ridge, at the main overlook, to become Hawk Ridge Nature Reserve. Duluth Audubon paid off the loan via fund-raising. Later, the City of Duluth and St. Louis County entered another agreement wherein the county transferred 200 acres of surrounding property, also tax-forfeited, to the city to serve as a buffer zone, and Duluth Audubon entered into a trust agreement with the City of Duluth, making the nonprofit the legal manager of Hawk Ridge. In 2004, that trust agreement was transferred to the Hawk Ridge Bird Observatory, which is responsible for protecting the habitat, maintaining the trails, ensuring an accurate count of migration, operating a field station where hawks and other birds are banded and released, and providing programs to educate school groups and the general public about hawks and their conservation.

Since 1991, when regular hours of counting were extended and count procedures standardized, the average annual count of raptors at the ridge has averaged eighty-two thousand. In 2003, Audubon Minnesota and the Minnesota Department of Natural Resources named Hawk Ridge the first Important Bird Area designated in Minnesota.

Weather-Dependent Migration

The magnitude of hawk migration on any day or week depends entirely on the weather. Hardly any birds fly during rain, and easterly winds aren't much more conducive to migration. Major flights virtually always happen on days when there is a westerly component to the wind; due west and northwest are best, but even southwest winds can be associated with an excellent flight.

As air warms, it expands, growing less dense. If the ground heats up more in one spot than an adjacent area, as it does along a shoreline or on pavement surrounded by vegetation, the expanding air rises as surrounding cooler, heavier air rushes in beneath. As the sun continues to beat, the rising air goes higher, forming what is called a thermal. If you look closely at a hot stove top or a dark car in direct sun, the rising heated air will be so agitated as it expands and rises that your eyes can actually see the small-scale thermal.

Hawks feel thermals as our heavy human bodies feel only the fiercest winds. Harriers and falcons have wings too narrow, and accipiters wings too short to bother with thermals except in ideal conditions. But in general, when hawks start migrating in the morning, they gravitate to areas where thermals are most likely to develop early, such as along a lakeshore or above a highway. When a hawk locates a thermal, it flies in a circle to maintain forward momentum as the rising air carries it upward in a spiral. Hawks don't need to feel rising air to locate thermals. When one hawk discovers a thermal, others notice its spiraling flight and join it, not out of any sense of sociability but because rising on a thermal and gliding to the next saves the energy of flapping on a long flight. Groups of hawks all spiraling on the same thermal are called kettles.

Those hawks that are most dependent on thermals may not start migrating until thermals grow strong in midmorning, and the first kettles may include only a few birds. But hour after hour, as more hawks are aloft, all gravitating toward the same thermals, a single kettle can number in the hundreds or, rarely, even the thousands. A large kettle can look like a tornado of hawks.

Wind affects not only the numbers of hawks flying but also how high they can rise in the sky. By afternoon on sunny days with light winds, thermals can carry hawks so high that they are impossible for us to see with the naked eye. Stronger winds break up the thermals, keeping the hawks lower.

Exceptional weather conditions are usually associated with exceptional numbers. Strong westerly winds were recorded for five straight days in September 1988 while a devastating fire raged in Yellowstone National Park. The Duluth sky grew a hazy yellowish brown, which the National Weather Service attributed to smoke from the distant blaze. On the smokiest day, September 9, a record thirty-two Swainson's Hawks were tallied at Hawk Ridge. This species breeds in the western states and provinces north to Alaska, and its normal migratory pathway is well west of Minnesota. Hawk Ridge averages only six each season, but this extraordinary number on a single day was apparently the result of these powerful winds pushing many birds east of their normal route.

In 2003, poor weather in the first two weeks of September held up Broad-winged Hawk migration as the birds waited until conditions were more favorable. On September 15, sunny skies and light west winds carried an unbelievable 101,698 Broad-winged Hawks over Hawk Ridge, more than double the previous high (47,919 on September 18, 1993). This new Hawk Ridge record may never be surpassed. Or will it?

The Challenges of Counting Hawks

Accurately counting enormous masses of hawks requires both expertise and experience. The counters at Hawk Ridge honed their skills by working with professional counters and volunteering at hawk lookouts for years before being hired for this count. On slow days, one counter may scan the skies all day without need of assistance. On large migration days, an assistant and additional volunteers help.

When only a handful of hawks are kettling, they can be counted without too much difficulty. But when kettles are large, counters don't even try to tally the swirling masses. The warm air rising in a thermal eventually reaches equilibrium with the surrounding atmosphere and stops rising. As hawks reach that height, they stream forward, all moving in the same direction toward the next thermal. Whether these birds are topping out in a narrow line or a wide band, they're all at the same altitude and moving at about the same speed and direction, and so are much easier to count. Because they are all headed in the same direction, trying to clear Lake Superior, it's usually not hard to keep track so birds are not counted twice.

Counting hawks is only half the job of a counter; accurately identifying each bird is equally crucial. Visitors often marvel at the counters' ability to identify what look like pepper specks in the sky. Accurate hawk identification takes careful study and lots of practical experience, but fortunately on days when hawks are kettling, the vast majority of those pepper specks are just one species.

On that amazing September day in 2003 when 101,698 Broad-winged Hawks were counted, the official counter tallied the ubiquitous broad-wings while assistants kept track of where the birds were moving so that birds moving from one kettle to another wouldn't be double counted. Meanwhile, other experienced hawk counters scanned all the kettles and the rest of the sky, picking out outliers. Identifying any bird is easier when it is near a known species, and the hundred thousand broad-wings filling the skies made it easy to detect the 631 raptors that were larger, smaller, or shaped differently that also passed through that day. How accurate are those numbers? Professional counters tend to be conservative in their numbers, whether they are counting by ones, tens, or higher numbers, and on the biggest days, the many

helpers serve as a check on overcounting. When so many birds fill the skies, the bigger danger is missing ones too high to see, or missing hawks moving by farther inland or out over the lake while the counter is focused on the huge numbers overhead. So the higher the number of hawks flying over, the easier it is to undercount.

Banding Hawks at Hawk Ridge

Counting hawks is only one of the avenues of research organized by the Hawk Ridge Bird Observatory. During migration, researchers at the banding station, in a

secluded area away from the main overlook, trap hawks. The birds are attracted to lures that look like easy prey, and get caught in nets. The researchers quickly retrieve them, place a uniquely numbered aluminum band provided by the U.S. Fish and Wildlife Service on one leg, and gather necessary data, such as species, sex, age (when possible to determine), and measurements. Birds are released within a few minutes of capture.

Over one hundred thousand raptors of twenty-three species (including owls) have been banded at Hawk Ridge since 1972. Only a small fraction are ever seen and reported again, but every one provides valuable information about migratory routes, longevity, and other important issues. The species captured in the largest numbers is the Sharp-shinned Hawk. Of the more than sixty thousand sharpies banded at Hawk Ridge over the years, less than 1 percent (388 as of this writing) have been recaptured at the banding station or found elsewhere. Yet from this small body of data we know that some of "our" sharpies have overwintered in the Midwest, while others have been found dead or recaptured in just about every Central American country and also in South America.

The banding station is operated from sunrise till sunset each day, and into the night when owls are migrating. During times when visitors are present, naturalists often bring banded birds to the main overlook to give visitors close-up looks before the birds are released. Birds are transported from the banding station in long tubes that immobilize and keep each hawk in the dark until it can be taken out, shown briefly, and released. A bird in the hand may not be worth two in the sky, but it's certainly easier to study and appreciate.

A Season of Hawks

Hawk migration begins almost imperceptibly in summer. During the first week of August, a hawk or two may be seen now and then, but some of these are local birds that aren't migrating yet. Hawk Ridge Bird Observatory's official counters start work on August 15. Turkey Vultures, Ospreys, Bald Eagles, Sharp-shinned Hawks, and American Kestrels are usually among the first tallied. Birders gather at the ridge in August to enjoy not just raptors but also migrating songbirds and hummingbirds and occasional spectacular late-afternoon flights of nighthawks. (Despite their name, Common Nighthawks are not raptors at all but, rather, relatives of the Whip-poor-will.) By the end of the month, daily raptor counts can be impressive—numbering a thousand or more on some days in late August.

The largest flights of many species peak in September, the month when total numbers of raptors and raptor watchers both reach their maximum. Although tens of thousands of hawks pass through in September, migration is very concentrated. Few days will have counts of more than a thousand, and it's quite possible to see twenty thousand hawks one day and only one or two the next. It's impossible to predict when the best days will occur without closely considering the weather, and forecasts can change dramatically between planning a trip and arriving at Hawk Ridge. Fortunately, on days with easterly winds, fog, drizzle, or rain, when few or no hawks are moving, Duluth birders have an alternative—movements of shorebirds, interesting gulls, and songbirds can be heavy along the lake in those conditions.

By October, some species have vanished even as numbers of other species are peaking. Bald and Golden Eagles and Red-tailed and Rough-legged Hawks continue to be abundant throughout the month. On the best days of November, Bald Eagles and Red-tailed Hawks may number in the hundreds, and many may still be passing through in December. The City of Duluth doesn't maintain a long stretch of the road to Hawk Ridge in winter, when it becomes part of a designated snowmobile trail with barriers to keep cars out. People planning a trip to Hawk Ridge in November or December should contact the Hawk Ridge office, at www. hawkridge.org, to find out if the road is open.

Visitors to Hawk Ridge are sometimes surprised to see a Great Horned Owl perched atop a pole near the counters' platform; it can take a moment or two to realize this is a plastic decoy. Why is it there? Great Horned Owls often kill roosting birds, including hawks, at night. So in daytime, many birds try to drive owls away. When a passing hawk sees the decoy, sometimes it dive-bombs it once or twice before realizing it's fake and moving on. This affords visitors some spectacular views of hawks in action.

Two-Way Passage

Hawks passing along Hawk Ridge are headed to wintering grounds. Those that survive the winter will be headed north in spring, and many will again be stopped by Lake Superior. Once they clear the western tip of the lake, they have no need to hug the North Shore, but a great many pass over West Duluth each spring. Hawk Ridge Bird Observatory sponsors an annual spring raptor count along West Skyline Parkway. The count begins in mid-March and continues through the end of May.

Unfortunately, there isn't as strong a migration-funneling effect at the tip of the lake as we enjoy at Hawk Ridge in fall, and depending on the wind direction, the count can take place at two different locations. Under most conditions, birds are counted just below Enger Tower at a pullout called Rice's Point. When the wind is east or northeast, the count moves to Thompson Hill.

For more information about the raptors' return journey or to join the counters, visit www.hawkridge.org.

Beyond Identification

Assigning a name to a bird is the only way you can meaningfully share your sighting with others and learn more about the species, but identification is merely the first step in experiencing hawks. Raptors have more power to stir human emotions than most birds. Their speed, agility, quick reactions, extraordinary vision, hunting prowess, and aggressiveness have inspired the names of countless sports teams. And of course a raptor was chosen by the Founding Fathers to serve as America's national emblem.

Symbols are not flesh and blood. As we sit on a rock observing migration, a low-flying falcon may meet our eyes, and we can't help but see the intelligence, the spirit, the unique being within. Each hawk's body structure, physiology, and natural history fascinate far beyond symbolism. Hawks do not exist to enrich our lives—they have lives of their own to lead. Henry Beston wrote in *The Outermost House*, "The animal shall not be measured by man. In a world older and more complete than ours they move finished and complete, gifted with extensions of the senses we have lost or never attained, living by voices we shall never hear. They are not brethren, they are not underlings; they are other nations, caught with ourselves in the net of life and time, fellow prisoners of the splendor and travail of the earth."

No book and, indeed, no library can capture all there is to know about hawks. Scientists and naturalists have uncovered some of their secrets, but we have barely scratched the surface. We hope this book whets your appetite, inspiring you to explore raptors more deeply. Hawk Ridge is the perfect starting place to experience a rich variety of these extraordinary creatures, both at close range in the hand and unfettered, flying wild and free through the sky on their timeless journeys. ◎◎

Vultures

(Order Accipitriformes, Family Cathartidae)

Vultures are among the most graceful of all flying creatures. Their bodies are surprisingly light for such large birds: a Turkey Vulture's wingspan is about six feet compared to a Bald Eagle's seven feet, yet the vulture weighs only 40 percent as much. Small wonder they can fly for many minutes without flapping once, and except during courtship virtually never flap more than ten times in succession.

Are vultures raptors? Ornithologists have been debating that question for centuries. On the one hand, like raptors they feed on meat, the only difference being that vultures seldom or never kill their own meals—Turkey Vultures prefer animals that are not just merely dead but really most sincerely dead. Vultures have a curved and hooked bill designed for tearing into carcasses. And vultures often migrate in kettles with hawks and eagles.

On the other hand, unlike hawks, vultures lack talons. Their claws are sharp, but their toes are not strong enough to kill or carry off food. They share several features with birds in the stork family, including bald heads and a curious habit of urinating on their legs to cool themselves. Vultures kettle on thermals, as do both hawks and storks. The American Ornithologists' Union (AOU) traditionally placed vultures in Falconiformes, the order that included hawks, eagles, and falcons. In 1998, the AOU moved the vultures to the order including storks, Ciconiiformes, bringing together the birds most associated with the beginning and the end of life. Sadly, the pleasingly ironic symmetry wasn't to last. In 2007, after new data cast the relationship between storks and vultures in doubt, the AOU returned the vultures to Falconiformes. In 2010, based on DNA evidence, they separated falcons from the rest of the hawks, placing eagles, hawks, and vultures in a new order they call Accipitriformes.

Two species of vultures have been reported at Hawk Ridge. The Turkey Vulture is very common here; the Black Vulture has been seen only once in the history of Hawk Ridge, on August 28, 2001.

Turkey Vulture *(Cathartes aura)*

A raptor floats ethereally overhead on outstretched black and silver wings raised in a shallow V. Its sheer beauty and gracefulness elicit gasps from a few Hawk Ridge visitors, who express disbelief when told that the beautiful bird is a Turkey Vulture. People often expect vultures to exude an unsavory aura, perhaps taking too seriously the aphorism, "you are what you eat." In northern Minnesota, Bald Eagles often dine on the same carcasses as vultures, yet no one expects eagles to be less than majestic.

"Bald" eagles have fully feathered heads; vultures are truly bald. When they thrust their face into a decomposing carcass, they have no head feathers to get gooped up. Laboratory tests and the experience of people who maintain Turkey Vultures in captivity indicate that the scavengers prefer fresh rather than spoiled meat. Unfortunately for them, they can't open thick hides. Animals killed in collisions with automobiles are often mangled enough for Turkey Vultures to eat. Otherwise, to feed on a large carcass, a vulture usually waits until it has been softened by decay or opened by mammalian scavengers. Turkey Vultures can easily get fresher fare from small animals with thin skin and apparently prefer small carcasses, unlike their stronger relative the Black Vulture.

Along shorelines, Turkey Vultures may feed on anything from bloated sea lion carcasses and decaying fish to piles of dead mayflies. As a pond dries in summer, vultures pick up dead and dying tadpoles and other aquatic species. They are adept at stripping skin off squirrels and opossums and can be surprisingly fastidious eaters: when devouring a dead skunk, they may even leave its scent glands intact. There are records of Turkey Vultures feeding on rotten pumpkins, but ornithologists assume that vultures feed directly on rotten fruit only when carrion is unavailable. Their regurgitated pellets contain a lot of plant material, but they probably ingest most of it indirectly, in the digestive tracts of the carcasses they eat. Meat itself is very digestible so is not found in pellets.

Unlike the other two North American vultures, the California Condor and the Black Vulture, Turkey Vultures have relatively huge olfactory centers in their brain and large nasal passages, giving them the ability to detect carcasses by smell. This allows them to find carrion hidden under forest canopy. But they also find food visually, both by noticing carcasses and by watching other vultures flying down to feed.

Scientists haven't determined exactly which odors Turkey Vultures detect to locate carrion. Decaying carcasses produce odoriferous gases, such as putrescine and cadaverine, which are probably attractive to vultures. Power companies add a foul-smelling aromatic hydrocarbon, ethyl mercaptan, to natural gas to help people detect leaks, and when a pipeline ruptures, circling vultures are often first to arrive on the scene. Putrescent carcasses are hotbeds of disease, but Turkey Vultures can eat them with impunity because their stomach secretions kill virtually all pathogens, including anthrax, botulinum, salmonella, and cholera bacteria.

Turkey Vultures are rather meek and mild, especially compared to Black Vultures, which often take over when they arrive at a carcass. Despite being sociable birds, kettling and roosting in large flocks, Turkey Vultures seldom feed in flocks. When more than one Turkey Vulture is present at a meal, all but one tend to draw back to wait their turns. A high percentage of a Turkey Vulture's diet is composed of small prey animals, possibly because those are easy to gobble down before other scavengers appear. Occasionally a vulture eats so much that it has difficulty taking off in flight.

Although very common and the most widely distributed of all New World vultures, ranging from southern Canada to the southern tip of South America, surprisingly little is known about the life history of Turkey Vultures. Researchers are prohibited from putting any kinds of bands on vulture legs because the birds excrete on their legs to regulate their temperature, and fecal material and urates accumulating on leg bands can cause dangerous lesions. Researchers tracking individual vultures can use wing tags, radio telemetry, and satellite transmitters, but surprisingly few in-depth studies have been made. Based on a few studies of individually marked birds, ornithologists believe they mate for life but may not stay together as a pair except during the breeding season.

Turkey Vultures do not build nests. They lay their eggs in dark recesses in ledges, caves, mammal burrows, hollowed-out logs, abandoned buildings, and other crevices, many inaccessible to humans. They have few defenses against predators except flight. When cornered, they may stomp their feet, hiss and grunt (they are one of the few birds that lack a syrinx to produce normal sounds), and vomit. If their food is repellant to us going in, it's reportedly even worse coming out. The Turkey Vulture's scientific name, *Cathartes,* is derived from the Latin for "purifier." It's probably just coincidence that vultures purify the landscape by devouring dead animals. The name was more likely applied to them in the sense of a cathartic or purgative, inspired by their unsavory but effective defense mechanism.

Turkey Vultures normally lay two eggs about a month after arriving on their nesting grounds in spring. The parents take turns incubating for about thirty or forty days; after hatching, the chicks are brooded continuously for another five days or so. The parents regurgitate well-digested food. The nestlings may thrust their bill into a parent's mouth to feed or may pick at a regurgitated mass on the nest floor. The chicks will make their first feeble flights at about sixty days of age and can fly well enough to reach high branches and land on distant rocks by the time they are seventy or eighty days old. When the chicks can follow their parents on foraging flights, the adults stop feeding them, and the chicks leave the nest area when they are about twelve weeks old.

Because Turkey Vultures are so adept at flying on thermals and updrafts, they tend to move about only when these are available. On windy mornings, they may leave their roosts to forage before sunrise, but most days their first flight is to a "post-roosting area," where numbers of vultures gather to preen or sunbathe until thermals develop, sometimes not until midmorning or later. They spend about a third of most days in flight, whether foraging for food, loafing in midair, or migrating.

In recent decades, the Turkey Vulture population has been expanding northward. Their numbers at Hawk Ridge have been stable.

Turkey Vultures at Hawk Ridge

Seasonal average of sightings
(over twenty years): 1,260

Earliest date of sighting: August 2

Latest date of sighting: November 26

Peak migration: late August through
October

Record daily count: 799 on
September 29, 1996

Record seasonal count: 1,952 in 1996

Black Vulture *(Coragyps atratus)*

As of early 2012, the Black Vulture has been recorded in Minnesota four times, and only once at Hawk Ridge, over a decade ago. The species range has been expanding northward in eastern states but so far does not seem to have undergone a similar expansion in the Midwest. Some ornithologists predict that this species will move farther north as climate warming continues, but this trend has not been noticeable at Hawk Ridge yet.

If a Black Vulture flew over, hawk watchers would recognize it by its head, which is larger than that of the Turkey Vulture and black at all ages; by its tail, which is much shorter and often spread wide in flight; and by its wings, which have whitish patches near the tips rather than along the trailing half. ◎◎

Black Vulture at Hawk Ridge

1 record: August 28, 2001

Ospreys

(Order Accipitriformes, Family Pandionidae)

Scientists set the Osprey apart from all other hawks in a family that includes just this species, found on every continent except Antarctica. Ospreys are the fishing specialists of the raptor world—virtually every scientist studying the species has found that at least 99 percent of an Osprey's diet is live-caught fish.

The Osprey is uniquely adapted for fishing. Its four toes are of equal length, and the outer of the three front toes is reversible, allowing an Osprey to grasp its fish evenly, the two normal front toes from both feet on one side of the fish and the hind toe and opposable toe on the other. No matter which way a fish thrashes, an Osprey's toes hold evenly so the fish can't pull out of the outer toes the way fish sometimes can when grasped by an eagle. To make that hold even more secure, the bottoms of Osprey toes are covered with raspy, backward-facing scales called spicules. In flight, Ospreys reduce wind resistance by thrusting the leg closer to the fish's head forward, thus taking advantage of the natural streamlining of a fish to fly aerodynamically.

Powerful wing strokes enable an Osprey to hover while scrutinizing the water below. Strength and the relatively long length of the inner wing allow it to rotate its wings to clear their tips from the water as it pulls both itself and a fish out of the water. Most of an Osprey's prey fish weigh between one-third and two-thirds of a pound. The heaviest fish verified by researchers weighed two and a half pounds, more than half the bird's body weight, and some observers insist that they've seen an Osprey lugging a four-pounder, which would be heavier than the bird itself.

Ospreys weigh about three and a half pounds—three-quarters the weight of Turkey Vultures. They have about the same wingspan, but Osprey wings bear a heavier load for their surface area, even when not burdened with a fish, because the wings are narrower.

Osprey *(Pandion haliaetus)*

A visitor arrives at Hawk Ridge just before sunrise to watch songbird migration. She knows that few raptors start moving that early, but when she faces east gazing at the opalescent sky, she spots a distant Osprey. During the next ten minutes, as she picks out various warblers feeding and coursing through the underbrush, the Osprey makes leisurely progress toward Hawk Ridge, finally passing directly over her head. In the next half hour, two more Ospreys follow the exact same path.

Every August and September, hundreds of Ospreys migrate over Hawk Ridge, one of the great environmental success stories of the twentieth century. Osprey numbers had plummeted while DDT was used during the 1950s and 1960s. When it was banned for agricultural use in the United States in 1972, the persistent pesticide and its dangerous metabolites slowly worked their way out of aquatic systems. As fish became safer for consumption, populations of the birds that ate them started to recover. Osprey numbers started increasing through the 1980s and more dramatically through the 1990s. The highest number ever counted in a single day at Hawk Ridge was 90, on September 17, 1997; that year's seasonal total counted from the main overlook was a record-breaking 568. Since then, Osprey numbers have dropped off, and the population seems to be declining slightly.

After DDT was banned in the United States, the Osprey's recovery took place fairly rapidly thanks to people setting out nest platforms. In Minnesota, the process was hastened when the Department of Natural Resources and the University of Minnesota's Raptor Center developed a reintroduction program in 1984. Bald Eagle populations bounced back much more quickly, probably because most Minnesota Bald Eagles winter within the United States; it took much longer for DDT to be banned in Central and South American countries, where most Ospreys spend half their lives. As eagles rebounded, they caused increasing problems for Ospreys. Eagles are known to eat Osprey eggs and chicks and to steal fish from adults, sometimes, though rarely, even killing adults during chases. When an eagle is spending time nearby, some pairs of Ospreys spend hours each day calling agitatedly. Even if eagles don't kill the young, an Osprey nest too close to an eagle nest has a good chance of failing because the Ospreys are so stressed by the larger birds' presence.

Great Horned Owls are another noteworthy predator of Ospreys, taking chicks and sometimes even killing and partially eating the attending adult at night. Raccoons take a significant toll too; where they are abundant, Ospreys usually build their nests over water to make them less accessible.

Ospreys arrive in northern Minnesota in early April and quickly start building their nest or repairing the

previous year's. They construct their large stick nests at the very top of live trees, snags, and power poles and readily build them on artificial platforms. After putting together a base of large sticks, they add a layer of smaller sticks and a variety of small materials, and on top of that a variety of fairly flat materials such as bark, mats of algae, and even paper and plastic bags. The dense, softer materials on top apparently prevent eggs from falling into cracks between the large sticks.

Pairs seem to mate for life, but it's hard for researchers to tease out whether the birds are bonded to each other or to the nest site—there is no evidence that they remain together off the breeding grounds. During their time together, they work as a team. They both build the nest, the male bringing more of the nesting materials, the female doing most of the arranging. The female incubates the eggs more than the male, while he provides virtually all the food for both of them. They move their feet about the eggs very delicately, with their toes and claws tightly closed. The clutch contains one to four eggs, usually three, which hatch in about thirty-seven days. The female broods the chicks almost constantly for the first two weeks after they hatch, and then intermittently after that. During very hot weather, she may stand over them providing shade without brooding them.

The male provides fish for the whole family. The female carefully breaks off small, soft pieces to feed the chicks, eating the bones, harder pieces, and tail herself.

When food is plentiful, all of the chicks are fed fairly evenly, there is little squabbling, and the chicks don't seem to have a dominance hierarchy. When there isn't enough food for everyone, the older nestlings dominate and get the lion's share; in harsh years, one or two of the smaller chicks may starve.

Ospreys are not into housekeeping. Adults and older chicks lean over the rim of the nest to defecate, but the droppings of small chicks, uneaten chunks of fish, and the bodies of any chicks that die remain in the nest. Common Grackles often build their nests among the sticks near the bottom of Osprey nests. The grackles may be able to retrieve fallen bits of fish and get some protection from the presence of the Ospreys. Ospreys don't seem to benefit from the presence of the grackles, but they don't seem to mind them either. Other birds that nest in active Osprey nests are Tree Swallows, European Starlings, and House Sparrows.

It's ironic that a bird that does so well nesting on man-made structures cannot adapt to captivity at all. Even experienced wildlife rehabilitators specializing in raptors have difficulty getting Ospreys to eat anything; this is probably due to both the species' skittish, high-strung nature and its uniquely narrow diet. Young Ospreys fledge at seven or eight weeks; their parents bring them fish for another two weeks as the young birds learn to catch them themselves, but when they stop getting handouts from their parents, they eat nothing but live-caught fish. It's possible that after this brief period of dependence, they no longer recognize dead fish as edible.

Young Ospreys that fledge with siblings stay together until migration. Fishing skills are apparently innate in Osprey. They do learn faster when watching their siblings, but when just one chick fledges, or in reintroduction projects when a single chick is released at a site, it can learn to fish entirely on its own.

Many Ospreys migrate all the way to South America. A Minnesota Osprey may fly more than 125,000 miles during its fifteen- to twenty-year lifetime. Ospreys probably don't expend much more energy migrating than they do on their daily foraging excursions. The ones wending their way over Hawk Ridge on exquisitely crooked wings will fly thousands of miles to their wintering grounds, and thousands of miles the following spring on their way back to us. The adventures they face during their long-distance travels must be left to our imaginations: they're not talking. ◎◎

Ospreys at Hawk Ridge

Seasonal average of sightings (over twenty years): 365

Earliest date of sighting: August 1

Latest date of sighting: November 6

Peak migration: late August through mid-October

Record daily count: 90 on September 17, 1997

Record seasonal count: 539 in 2005

Eagles

(Order Accipitriformes, Family Accipitridae)

Eagles may be among the most recognizable of all birds, but to an ornithologist, "eagle" is just a name, not the designation of a natural grouping of species. The Bald Eagle belongs to a genus of fishing eagles, *Haliaeetus,* while the unrelated Golden Eagle, in *Aquila,* is separated by ten species in the Minnesota Ornithologists' Union checklist. The Bald Eagle's uncompromising visage serves as our national emblem; the Golden Eagle inspired both Alfred, Lord Tennyson's famous poem and the description of curved noses as "aquiline."

The two species share the distinction of being the heaviest of Minnesota's birds of prey, weighing about eleven or twelve pounds. The Golden Eagle's wingspan is seven feet, the Bald Eagle's just a few inches less. Young Bald Eagles are often mistaken for goldens before they assume white feathers on their head and tail at four or five years of age, but their differences involve more than just feather color. Bald Eagles have unfeathered legs, presumably to reduce drag when fishing, while Golden Eagle legs are fairly thickly feathered down to their toes. Bald Eagles are scavengers year-round, especially in winter. Golden Eagles are more like Red-tailed Hawks and other buteos, superbly skilled hunters of mammalian prey.

Bald Eagles are far more abundant than Golden Eagles at Hawk Ridge, averaging about 3,000 each fall compared to about 125 goldens. Bald Eagles start migrating through in mid-August; some of the ones that fly over in August and September may be southern birds that had wended their way north after their breeding season. The largest migrations of Bald Eagles tend to be in late October and November. On the biggest eagle migration day recorded at Hawk Ridge, November 22, 1994, a thrilling 743 Bald Eagles flew over. Golden Eagles also peak in October and November; the peak day for goldens was Halloween 2008, when 53 flew by.

Based on Hawk Ridge counts during the past twenty years, numbers of both Bald and Golden Eagles have been steadily increasing.

Bald Eagle *(Haliaeetus leucocephalus)*

From a distance, the first thing we notice is hugeness. The white head and tail can disappear into blue sky, leaving what appears to be a flying board, flat and solid, unwavering in the wind. As it draws closer, we see the primary wing feathers jutting out like fingertips, and as it banks, the gleaming white head and tail pop into visibility. As it reaches the main overlook, hawk watchers tracking it with binoculars get a clear view of the forbidding bill and uncompromising eyes.

When helping design the national emblem, Ben Franklin ignored the commanding aspect of the Bald Eagle and focused on its scavenging habits, how it steals fish from Ospreys, and how "like those among Men who live by Sharping and Robbing he is generally poor and often very lousy." Franklin preferred the turkey, a bird more emblematic of the nation's wealth in natural resources.

Choosing the eagle was a committee decision. Thomas Jefferson and John Adams overruled Franklin—they wanted a powerful symbol of the new nation's military might and will and thought the eagle fit the bill. Franklin apparently disagreed: he noted about the eagle in his famous letter to his daughter, "Besides he is a rank Coward: The little King Bird not bigger than a Sparrow attacks him boldly and drives him out of the District. He is therefore by no means a proper Emblem for the brave and honest Cincinnati of America who have driven all the King birds from our Country." Eastern Kingbirds do indeed dive-bomb any eagle that approaches their territory, repeatedly striking the huge raptor's head and back so aggressively that the eagle may well turn off course to avoid them. But kingbirds attack turkeys with the same success—these diminutive tyrants come by their scientific name *Tyrannus tyrannus* honestly. For wise eagles, the better part of valor is discretion, shown by simply avoiding the tiny pugilists.

Bald Eagles may be emblematic of patriotism in America, but they haven't been treated with much respect over their history. Alaska established a fifty-cent bounty on Bald Eagles in 1917 and increased the bounty to two dollars in 1949. Over 128,000 bounties were paid out for dead Bald Eagles between 1917 and 1952. The Bald Eagle Protection Act took effect in 1940, and the federal government overruled the bounty on eagles in 1952.

Unfortunately, just as eagles started receiving some protection from outright shooting, a more insidious threat began spreading over the continent and beyond, accumulating in the fatty tissue of fish and concentrating in the fat of animals that ate them. DDT, first used in the American landscape in 1947, probably killed few if any Bald Eagles directly, because eagles don't metabolize huge amounts of stored fat during single migration flights as songbirds do, pouring lethal levels into their bloodstream all at once. But the buildup of DDT and its metabolite DDE in their tissues caused female eagles to produce eggs with dangerously thin shells that usually broke during incubation. In Florida in 1958, only three Bald Eagle nests were successful compared to about one hundred a decade earlier. It takes four or five years for Bald Eagles to molt into adult feathers, and by the 1970s, virtually all the eagles seen at Hawk Ridge and other places were adults, because so few eagle chicks had been produced in previous years.

DDT was banned for most uses in the United States in 1972. Thanks to that ban and the passage of the Endangered Species Act in 1973, the Bald Eagle is now thriving all over North America. By the 1980s, eagles were breeding in Florida with about the same success as before the DDT years.

Although wild Bald Eagles are reproducing successfully and individuals can now lead healthy lives in the wild for decades, eagles are still dying in unacceptable numbers due to one tragically preventable cause: lead poisoning. Eagles feed on gut piles and unretrieved carcasses of deer, pheasants, and other game animals that harbor shot or bullet fragments. The University of Minnesota's Raptor Center conducted a thirteen-year study, from 1996 to 2009, that concluded that spent ammunition is an important source of lead exposure for eagles.

Nontoxic shot has been required for all small-game hunting on federal waterfowl production areas in Minnesota since 1998 but is still allowed on most upland areas. The Minnesota Department of Natural Resources made a recommendation to the legislature in 2008 to ban toxic shot for upland game on all state-managed lands, as twenty-three other states do, but so far it is banned on just forty-five acres managed for dove hunting.

Despite the serious issue of lead poisoning, eagles have become abundant enough that a species once associated with genuine wilderness has grown increasingly tolerant of people. One pair nests in a large pine tree near the parking lot of the large Duluth high school in the residential neighborhood beneath Hawk Ridge.

Thanks to the many research projects started after the Endangered Species Act took effect, a great many more eagles were banded in the 1970s and afterward than had ever been banded before; recoveries of these birds in the past decade have been setting new longevity records for the species. Every Bald Eagle known to have survived in the wild for longer than twenty-two years was banded in 1970 or later; some have lived more than thirty-one years. In captivity, they may live even longer. In 2012, Marge Gibson, executive director of the Raptor Education Group in Wisconsin, still had in her care a female Bald Eagle that she had rescued while serving as director of the Bald Eagle response team in the aftermath of the *Exxon Valdez* oil spill in Alaska in 1989. The bird was not releasable and has served as an education bird and foster mother for young eagles at the center. At the time Gibson rescued her, the eagle was at least twenty-nine years old, making her at least fifty-two years old in 2012.

Bald Eagles are believed to mate for life, but this is mostly based on circumstantial evidence because so few eagle pairs have both been banded or permanently marked. We don't know if pairs remain together over the winter or if they take separate vacations. If it's true that they do mate for life, we don't even know if they form a permanent genuine bond or if they simply can't work out a property settlement—both birds seem very focused on maintaining the nest, both before and after the nesting season.

It may take weeks or even months for a pair of Bald Eagles to construct a new nest, though at least one was built in just four days. A pair selects one of the largest trees available with limbs strong enough to bear the weight of a large nest. In Minnesota the preferred species is the white pine; the eagles usually build in the top quarter of the tree just below the crown. They pick up sticks from the ground and break some off nearby trees to form the main structure, then fill in the center with twigs, grasses, mosses, and other small materials, and finally add even finer, softer materials, including their own downy feathers. They continue adding materials throughout the nesting season.

Nest building in spring may help synchronize a pair's hormonal levels so they will be physiologically ready for breeding at the same time. When adults with established nests return to their territory in spring, they make repairs at the same time that they are occasionally engaging in their spectacular courtship flights. Females produce two eggs in 79 percent of all clutches; single-egg clutches are more common than those with three eggs. The eggs are laid three or four days apart. The female does the majority of incubating, but the male helps. It takes about thirty-five days for the eggs to hatch; it can take a full day for a chick to pip the egg and emerge. The first egg to be laid hatches three or four days before the next.

The pink skin of newly hatched eaglets is covered with light gray down. One or both parents attend the nest for the first month, when the young are virtually never left alone. Both parents brood the chicks, hunt, and feed them, though for the first two weeks the female does most of the brooding, and the male most of the hunting. The chicks grow rapidly, often outweighing their parents by the time they fledge at eight to fourteen weeks, though they continue to develop muscle mass, and their flight feathers continue to grow after they fledge.

The young remain with their parents for up to six more weeks after they fledge, depending on their parents for food as they practice hunting. They seem to develop their skills by trial and error and usually pick up dead fish along shorelines and floating on the water before they learn to catch live ones.

Within a couple of months of their first flights, young eagles are off on their own. It's a big, beautiful, dangerous world out there, but their increasing numbers at Hawk Ridge are proof that plenty of young Bald Eagles are finding their place and thriving.

Bald Eagles at Hawk Ridge

Seasonal average of sightings (over twenty years): 3,094
Earliest date of sighting: August 13
Latest date of sighting: December 13
Peak migration: September through November
Record daily count: 743 on November 22, 1994
Record seasonal count: 5,725 in 2011

Golden Eagle *(Aquila chrysaetos)*

October days may be uncomfortably cold at Hawk Ridge. On this day, temperatures haven't warmed much after the season's first hard freeze the night before, and blustery northwest winds sharpen the chill. The brilliant blue sky is dotted with big cumulus clouds, heightening the intensity of the fall colors. When hawk watchers can draw their eyes from the skies to look at the neighborhood below, they marvel at the patchwork of yellows, oranges, and crimsons ending abruptly at Lake Superior, sparkling with wave action. The beautiful day, high-pressure system, and the hundreds of Bald Eagles and Red-tailed Hawks aloft in the clear skies buoy the hawk watchers' spirits.

Hawk watchers spot a huge, dark raptor off in the distance. It could easily be passed off as a Bald Eagle from this distance, but the counters notice something curious: the wings are held at a slight uptilt. They switch to their spotting scopes and don't allow their eyes to wander off the bird for even a second as it draws closer. It's still way off in the distance when it banks, and they detect the well-defined whitish patches at the base of the primary wing feathers and the dark terminal band on the white tail that confirm their suspicions. One shouts out, "Golden Eagle!" The count interpreter points out the bird to dozens of visitors and explains how the counters made the identification. A few, new at hawk watching, feel bewildered. To their eyes it looks exactly like an immature Bald Eagle from so far away. But the count interpreter keeps them on the bird as it slowly draws closer, reminding them how splotchy the whitish areas on young Bald Eagles are. The three patches of white on a young golden are much more clearly defined. This particular Golden Eagle is exceptionally cooperative, slowly but surely making its way directly overhead, and a young Bald Eagle wings by just a minute later to help clarify the field marks that distinguish the two.

Like Bald Eagles, the steadily increasing numbers of Golden Eagles counted at Hawk Ridge are a conservation success story. Golden Eagles were exposed to DDT but not at the concentrations that birds feeding on large fish at the top of the aquatic food chain were, so pesticides weren't as serious a problem for them. But people have persecuted Golden Eagles for many centuries. Europeans trapped, shot, and poisoned them. Some Hopi people in Arizona continue to legally capture and kill, for ritualistic sacrifice, Golden Eagles within the Wupatki National Monument. And many people have illegally shot and trapped Golden Eagles, sometimes for feathers, sometimes out of frustration at their killing livestock, and sometimes, ironically, because some

hunters and trappers do not want eagles to do for survival what they themselves do for sport or profit.

Golden Eagles are superb hunters. Alfred, Lord Tennyson's famous poem "The Eagle" from 1851 conjures both the wildness of the lands where Golden Eagles live and their ability to drop on prey with astonishing speed:

> He clasps the crag with crooked hands;
> Close to the sun in lonely lands,
> Ringed with the azure world, he stands.
>
> The wrinkled sea beneath him crawls;
> He watches from his mountain walls,
> And like a thunderbolt he falls.

In America, Golden Eagles kill very large prey, including cranes, small deer and antelopes, and young domestic livestock, but they live primarily on hares, rabbits, ground squirrels, and prairie dogs. Most often they specialize on jackrabbits. In the West, Golden Eagle reproductive rates fluctuate with black-tailed jackrabbit population cycles.

Golden Eagles don't nest in Minnesota. Some of those seen at Hawk Ridge may have flown in from the West, from the Plains states and provinces and through the Rocky Mountains and Alaska. Transmitter studies of Golden Eagles wintering in Minnesota by Mark Martell

and others have shown that some goldens head north to the subarctic around Hudson Bay for the summer. With greater education and enforcement of wildlife protection laws, their breeding success has improved over much of their range. Populations in the Southwest are still in jeopardy.

Like Bald Eagles, Golden Eagles don't begin breeding until at least five years of age. Each pair usually remains on their large breeding territory, ranging in size from seven to twelve square miles, from the time they arrive in spring until they migrate in fall, defending the territory from other Golden Eagles. Nest building and maintenance are essential components of Golden Eagle courtship. Some pairs may build as many as fourteen nests on their territory over time, though they use only one during a season. They usually produce two eggs, but normally only one chick will survive to fledge.

Every sighting of a Golden Eagle at Hawk Ridge arouses a level of excitement that can't be explained simply by the rarity of this species, so outnumbered by Bald Eagles. It's not that we normally get amazing views of Golden Eagles; only in the best light and at reasonably close range do we see the golden brown head and neck feathers on adults that give the species its name, and usually both adults and immatures are seen at a distance rather than directly over the main overlook. Yet this magnificent bird, which most closely fits the descriptions of the "thunder birds" of Native American folklore and Tennyson's thunderbolt, has a powerful pull on our imaginations, and our thirst to see one is quenched by even a poor glimpse. When one does fly nearer, the thrill is tangible. ◎◎

Golden Eagles at Hawk Ridge

Seasonal average of sightings (over twenty years): 129

Earliest date of sighting: September 6

Latest date of sighting: December 12

Peak migration: October to November

Record daily count: 53 on October 31, 2008

Record seasonal count: 223 in 2009

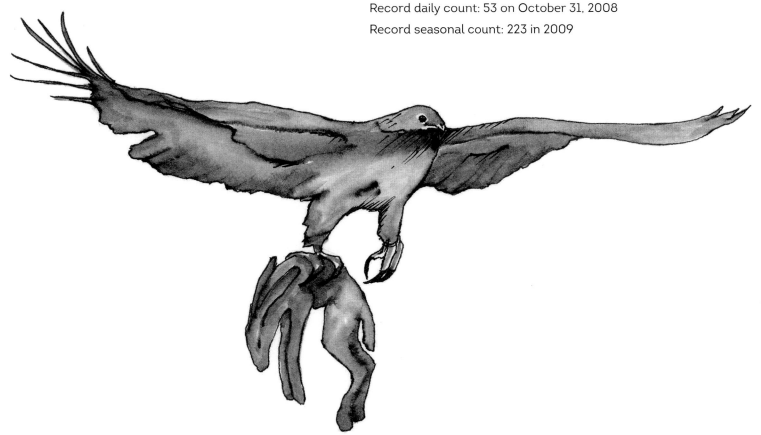

Harriers and Kites

(Order Accipitriformes, Family Accipitridae)

At Hawk Ridge, the rather common hawk with long, narrow but non-pointed wings and a long, narrow tail is the Northern Harrier, which has a light, buoyant flight often described as butterfly-like. The hawk occasionally confused with it is the Mississippi Kite, which has been reported from Hawk Ridge twelve times since 1972.

Harriers and kites are not closely related but share superficial characteristics in their overall shape and buoyant flight, at least when the kite is leisurely hawking for insects. Mississippi Kites can also fly with considerable speed, appearing falcon-like. The kite is much smaller than the harrier, with a thirty-five-inch wingspan compared to the harrier's forty-four inches, but size can be hard to determine from a distance unless the bird is near one of known size. The harrier has a bold white rump, the kite an all-black tail. And the kite is far, far less likely to appear at Hawk Ridge.

Both the harrier and kite are the only North American representatives of their respective genera. The Northern Harrier belongs to *Circus,* a word coined by ancient Romans and Greeks to apply to hawks in general after observing that many hawks circle in the sky. Ironically, Linnaeus applied the name to one of the few hawks that virtually never circle. The Mississippi Kite belongs to *Ictinia,* from a Greek word for kites that may be etymologically related to Icarus. Fortunately, the bird's wings are not fastened with wax.

Harrier numbers tend to fluctuate in cycles of four to five years, probably due to their dependence on voles as a prey species. Throughout the continent, this species has been on the National Audubon Society's Blue List of declining species and is listed as endangered in Iowa and threatened in Wisconsin. Hawk Ridge counts of harriers have been fairly stable over the past twenty years, but Hawk Ridge numbers reflect the breeding populations west and north of us, away from where harriers are facing the worst problems.

We don't have enough data from Mississippi Kites to detect any trends at Hawk Ridge; their breeding range is too far south of Minnesota to expect any changes in the foreseeable future.

Northern Harrier *(Circus cyaneus)*

Hawk watchers come to Hawk Ridge to see as many hawks as possible, and so they spend much of their time facing east-northeast in search of birds coming toward the main overlook from over Moose Mountain, a ridge running perpendicular to the lake two and a half miles up the shore. Birders also face south to cover the area between the ridge and the lake, watching for birds flying along the shoreline. Experienced birders know that they must also scan inland if they want to spot Northern Harriers, one of the prettiest hawks that fly along the ridge. Adult male harriers have the gray and white body and black wing tip pattern of a gull. That lovely plumage is highlighted by gracefully loose wing strokes and buoyant, tilting glides, which make these "gray ghosts" a truly thrilling sight. Adult females are brown above and pale beneath, heavily streaked with brown. In autumn, the young of the year are a richer brown than adult females, with deep cinnamon faces and wing linings; this color fades by the following spring. All harriers have a large white rump patch, conspicuous from most angles.

Northern Harriers are creatures of open country, nesting in marshes and pastures, their populations densest and most successful in undisturbed habitat. They feed heavily on rodents, especially meadow voles, and also take many small and medium-sized birds, reptiles, and frogs. Males have slightly shorter wings and a lighter body, making them more agile and maneuverable in flight than females. This may explain why males catch more birds than females do. While hunting, harriers often glide low over fields and marshes, looking down as they wing through. They are noted for hovering along the lead line of grass fires, darting down to grab animals trying to escape. A group of Minnesota birders on a visit to the Aransas National Wildlife Refuge in Texas once watched a harrier following a coyote through the marsh. Whenever the coyote flushed a rail or small heron, the harrier would dart at the smaller bird. The group watched for five minutes and saw several attempts but no successful catches.

Northern Harriers are unique among raptors in that many of the females join together to form a "harem," all mating with the same male. This polygynous system seems based on females strongly preferring to mate with very high-quality hunters; no yearling males are ever selected as mates by a harem. In Wisconsin, about 11 to 14 percent of males and about 20 to 29 percent of females mate in these polygynous systems; other adult birds may be monogamous, even in the same areas. The number of polygynous groups seems to increase in years when voles are most abundant, possibly maximizing the numbers of young produced in those years, contributing to the harrier's cyclically fluctuating population.

From about October through spring migration, Northern Harriers are fairly sociable, sometimes gathering in large communal roosts, which may also include Short-eared Owls. During some years, large numbers winter in Minnesota, and occasionally several may spend their days hunting in the same or adjacent fields.

Harriers are ground nesters. After pairs and breeding harems form, the female gathers most of the nest material and does most of the construction. She also does all of the incubating, although males will occasionally shade the three to five eggs for a few minutes when the female is off the nest. It takes about a month for the eggs to hatch. Females brood the chicks on and off during the day, more when the temperature is cool or during rain than when it's more pleasant. She also often stands with outspread wings to shade or shelter nestlings during rain and in early afternoon when the sun is high.

The young back off the nest to defecate. Females eat or remove leftover food, skeletal remains of prey, and any pellets the young regurgitate. If a female dies, these

items accumulate in the nest. If a chick falls from the nest while tiny, the female retrieves it, carefully carrying it by the nape in her bill. By the time chicks are two weeks old or so, they start moving into the vegetation surrounding the ground nest and begin using the nest just for feeding, brooding, and roosting. Harriers start making their first flights when about four or five weeks old. By this time, the parents are roosting away from the nest area and returning only for feedings. When the young can fly, parents give them food on the wing in aerial exchanges. Families break up when the chicks are about seven or eight weeks old, and the chicks don't appear to remain together.

Young harriers tend to migrate before adults, and among adults, females tend to move before males. So visitors to Hawk Ridge can expect to see mostly immature birds and adult females in September and the largest flights of the gray ghosts in October.

Northern Harriers at Hawk Ridge

Seasonal average of sightings (over twenty years): 515

Earliest date of sighting: August 2

Latest date of sighting: November 28

Peak migration: September and October

Record daily count: 216 on September 17, 1994

Record seasonal count: 1,100 in 1999

Mississippi Kite (*Ictinia mississippiensis*)

On twelve days since 1972, lucky hawk watchers at Hawk Ridge have witnessed a small miracle: a Mississippi Kite darting through the sky catching dragonflies. This lovely little hawk, falcon-like in shape, belongs a thousand miles south of here. No one understands what impulse has carried a dozen of these birds to Duluth over the years, but the birds have always arrived on one of the days in late August and early September when the skies along Lake Superior are dotted with dragonflies migrating along the same course as hawks. One kite spent more than an hour at the main overlook, darting through the sky in a breathtaking display as it snatched up dragonflies and then fed on them on the wing.

The Mississippi Kite does not seem to be expanding its range. No one knows why a handful of individuals have wandered our way, and no one can predict the next visit of one of these pretty little hawks. But that's one of the joys of hawk watching: when you scan the skies, you never know what's going to turn up next. ◎◎

Mississippi Kites at Hawk Ridge

12 records since 1991

Earliest date of sighting: August 30

Latest date of sighting: September 15

Record seasonal count: 3 in 2004

Accipiters

(Order Accipitriformes, Family Accipitridae)

The three American accipiters range in size from the fairly tiny male Sharp-shinned Hawk, about the size of a Blue Jay, to the huge female Northern Goshawk, which can be larger than ravens. Regardless of size, hawk watchers recognize accipiters by their characteristic shape: short, rounded wings and long, narrow tail. Accipiters can be seen at Hawk Ridge virtually every day during migration, usually much lower and closer than other raptors because they hunt at treetop height as they migrate along. The birds accipiters feed on are also abundant during accipiter migration, so these hawks often have a large bulge on one side of their lower neck—a distended crop. The crop is an offshoot of the esophagus where a bolus of food can be stored en route to the stomach. It may be swollen with food for an hour or two after a hawk feeds. Seeing this on a flying bird takes practice, but when an accipiter with a distended crop is brought to the main overlook from the banding station, the naturalists point it out. The accipiter habit of actively hunting while migrating is also why they are caught in such large numbers at the banding station—they are the ones most easily attracted to the nets by lures that look like injured birds.

All accipiters are specially adapted for taking birds by surprise and snatching them out of the air. Their long tail provides excellent maneuverability, and their short wings are designed for quick takeoffs and rapid flight through forests. They have huge, fairly forward-facing eyes for spotting and tracking prey, and their very long middle toe and long, sharp claws give them a bigger margin of error in grabbing flying birds. Thick, rough-textured pads beneath their toes cushion the joints and exert pressure to more firmly grasp struggling prey.

Females are larger than males in virtually every raptor species, but this difference is most extreme in accipiters. Having two sizes within a single pair allows one to specialize on quick-flying smaller birds as the other focuses on slower but heavier ones; between the two, they can capitalize on a wider range of prey species within their territory. Why is the female the larger bird? Bird students have been debating this for centuries.

Because both birds incubate the eggs and defend the nest, there is no reason for the male to be larger. The female's greater mass provides her more resources for egg production, and any weight fluctuations during egg production are proportionally smaller. Because the female does spend more time at the nest, she's the one more often defending the young. Finally, those excellent bird-killing tools that are standard equipment on an accipiter have to be held in check during mating; this is easier when it's the larger, more powerful female in the vulnerable position.

Accipiter plumage changes dramatically between a bird's first year and its second, and during the following years, eye color changes from bright yellow to orange to deep ruby red, making it easy to figure out the approximate age of an individual accipiter. These changes are not as precise a measure as, say, tree rings, but are still a handy clue for hawk banders trying to determine the age of a bird. They are even more essential for the hawks themselves. Healthy birds grow increasingly attractive to the opposite sex as they grow older, their age solid proof that they are excellent hunters and likely to have all the skills necessary for feeding and protecting young.

Three species of accipiters migrate through Hawk Ridge. The diminutive Sharp-shinned Hawk tends to be the most numerous hawk from day to day throughout the season, outnumbered by other hawks only under ideal weather conditions for soaring species. It is also the species caught in greatest numbers at the banding station. Migration of the largest, the Northern Goshawk, peaks later in the season and is characterized by dramatic cyclic fluctuations in numbers from year to year. Cooper's Hawk, the most familiar backyard hawk throughout much of the United States, is the least common of the three accipiters at Hawk Ridge.

Sharp-shinned Hawk *(Accipiter striatus)*

A bird about the size of a Blue Jay flies over the main overlook, just a foot or two above the heads of the tallest hawk watchers. A few beginners, expecting hawks to be large, are surprised to hear more experienced birders calling out, "Another sharpie!" and "Here comes a shin!"

The diminutive Sharp-shinned Hawk may be small, but it packs a wallop. Although most of its prey species are the size of sparrows and warblers, sharpies have been documented killing, carrying off, and eating robins and even killing flickers.

Sharp-shinned Hawks often dive-bomb flying or roosting Blue Jays and have been observed killing jays at least a few times at Hawk Ridge, but how often they strike at and how often they kill them isn't known. In most raptors, including both hawks and owls, females are larger than males, but in no species is the size difference as pronounced as in Sharp-shinned Hawks—the average weight of males averages just 57 percent of the body mass of females. It's likely that most of the larger prey species are taken by female sharpies.

At Hawk Ridge, only two raptors are regularly seen actively hunting at the main overlook. American Kestrels pluck dragonflies out of the sky during the big dragonfly movements on nice days in August and early September, and Sharp-shinned Hawks occasionally snatch up little birds as they fly over at treetop height. Sharp-shinned Hawks seem most effective at grabbing birds right after the birds flush in a panic. If birds elude a sharpie in the first critical moments, the little hawk usually gives up and moves on to try elsewhere. If the hawk does succeed in capturing a bird, it carries it to a tall branch to pluck its feathers and eat it. Sharpies are often seen in little groups of two and three. It's possible that the birds aren't cooperating so much as exploiting one another, because as the first one

startles up small birds, they may be so busy eluding that one that they don't immediately notice another hawk right there.

Many field guides describe accipiter flight as "flap-flap-glide." This is a useful description of a Sharp-shinned Hawk's typical hunting flights. As they migrate, however, sharpies often rise on headwinds, opening their wings and gliding for many seconds. And they are often seen in kettles with other hawks, sometimes floating on open wings, tail spread, without flapping at all for a minute or longer. Hawk watchers suspect that after an individual Sharp-shinned Hawk has eaten its fill, it takes advantage of the same method of covering large distances with a minimum of work used by so many other hawks, especially when thermals are strong.

Since migration doesn't limit their ability to hunt except in heavy rain, Sharp-shinned Hawks are more likely to fly over Hawk Ridge on east-wind and drizzly days than other raptors. On perfect migration days, when a great many songbirds are also moving along the ridge, sharpies are usually seen at much closer range than most of the other species, taking advantage of the abundant prey.

On average, first-year sharpies migrate earlier in the season than adults, and in both age classes, males earlier than females. Based on banding returns from many migration spots, adult females migrate shorter distances than adult males and all immatures, most likely because adult females can dominate all other groups in the use of an area's resources. Migration is hazardous and energy intensive, so it makes sense that the most dominant birds would migrate as short a distance as possible, forcing the others to travel farther.

The body of data collected at major migration sites such as Hawk Ridge is essential for discovering population trends for some secretive raptors like Sharp-shinned Hawks. Instruments such as the Breeding Bird Survey, a long-term annual roadside count coordinated by the U.S. Geological Survey's Patuxent Wildlife Research Center and Environment Canada's Canadian Wildlife Service, don't provide a complete picture, but no single set of data, including hawk watches, can. Sharp-shinned Hawk numbers at Hawk Ridge and some other count sites in the East have been dropping a bit over the past twenty years, but because hawk migration from year to year is so affected by weather, the trend isn't statistically significant. It's possible that warmer winters and more bird-feeding stations (sharpies often spend winter near bird feeders to capitalize on easy prey) are allowing more Sharp-shinned Hawks to winter farther north, causing a real decrease in numbers at migration hot spots that might not indicate a decrease in the overall population. We can't know exactly what is going on with them, but right now Sharp-shinned Hawks appear to be doing well.

When Sharp-shinned Hawks are not migrating, they tend to be secretive and solitary. Even during the nesting season, pairs don't seem particularly bonded, spending as much time apart as possible except during the dramatic courtship flights that precede mating. Both sexes bring the nest materials, but the female does most or all the construction of the relatively large stick nest built fairly high and deep in a thick tree, often a conifer. She is probably the only parent who incubates the eggs. Although eggs are laid on alternate days, the four or five chicks generally hatch out within forty-eight hours of each other. Being so much tinier, the males develop more quickly than the females. The mother broods the young until they are sixteen to twenty-three days old, while the father provides all the food for both her and the chicks until brooding is done. Then both parents hunt and deliver food, though usually the female is the one who tears items apart for the chicks.

When the young hawks are about six weeks old, the parents start cutting back on feeding, even though prey is usually very abundant. As the young start flying, the parents start delivering food to them in midair. Sometimes fledglings from other nests compete for these feedings, and the parents don't seem to recognize that those are intruders. Fledglings remain dependent on parents for three or four weeks and begin migrating before the adults. Soon they are winging their way along Hawk Ridge, filling hawk watchers with delight as they glide along the treacherous path toward adulthood.

Sharp-shinned Hawks at Hawk Ridge
Seasonal average of sightings (over twenty years): 16,033
Earliest date of sighting: August 5
Latest date of sighting: December 8

Peak migration: September and October
Record daily count: 2,040 on October 8, 2003
Record seasonal count: 21,352 in 1997

Cooper's Hawk (Accipiter cooperii)
When hawk watchers hear someone yell "Coop!" they instantly search the skies. Cooper's Hawks are easy to find in many suburban and even urban neighborhoods in the eastern United States, but they are by far the least abundant of the accipiters found at Hawk Ridge.

Cooper's Hawks superficially look very similar to sharp-shins, both in first-year and adult plumage, and share the same eye-color transition from yellow to orange to red. And they share the accipiter flight silhouette—short, broad wings and long, narrow tail. But in flight, sharpies have a somewhat hunched appearance due to the curved leading edge of their wings; Cooper's Hawks hold their wings straighter, with their proportionally thicker neck and head jutting out more. Some hawk watchers describe the overall shape differences as a letter T for sharpies and a cross for coops. Cooper's Hawk tail feathers taper, giving the tail a more rounded appearance, enhanced by a greater amount of white at the very tip. Cooper's Hawks are significantly larger than sharpies, with no overlap between male coops and female sharpies, but size is virtually impossible to judge accurately unless birds of known size are nearby for comparison.

Cooper's Hawks have been increasing in many areas since the 1990s, perhaps as the species has become adapted to nesting in urban areas. Many nature centers and bird clubs are bombarded in spring and summer with phone calls from people distressed that Cooper's Hawks are killing the birds at their feeders. When Cooper's Hawks are nesting near enough to become regular backyard hunters, it's wise to close down bird feeders while the raptors are present. When a Cooper's Hawk flies through a yard, very often feeder birds fly away in such a panic that one or two may strike a window and be killed in addition to any the hawk captures. And sometimes a Cooper's Hawk may be so intent on a bird it is chasing that it doesn't notice the window itself. If you do continue feeding birds when a Cooper's Hawk is nesting or wintering nearby, it's a mercy to make feeders facing your windows more bird safe. The American Bird Conservancy offers excellent suggestions for making windows more visible to birds on their website at www.abcbirds.org.

In the Midwest, Cooper's Hawks begin pairing up and breeding in March. Some individuals select the same mate from previous years, while others do not. Larger birds tend to choose mates that are larger too,

and size appears to have a genetic component. When young birds enter their second year, they are still wearing immature plumage, a clear indication of their inexperience, making these birds far less desirable as mates, but a handful do manage to attract a mate when just a year old.

Males do most of the work constructing a nest, but between bouts of building, they don't stay near the nest very long. Cooper's Hawks produce an average of three to five eggs in a clutch but occasionally as many as seven or as few as one. Larger pairs produce more chicks than smaller pairs do, and Cooper's Hawks nesting in urban areas tend to produce more young than do those nesting in more rural areas. Incubation usually begins after the third egg is laid, so the first three chicks hatch at close to the same time and will be larger than any siblings hatching later. Females incubate most of the time during day and all night. At least one female incubated 84 percent of daylight hours. Males incubate in bouts of ten to twenty-five minutes two or three times a day. Males do most or all of the hunting at first. When the female is on or near the nest, the male delivers food to her at a nearby perch, and she brings it to the chicks, tearing it into appropriately sized pieces for their ages. If the female is away when the male arrives with prey, he places it in the nest but doesn't feed the chicks.

Cooper's Hawk chicks fledge at about a month of age. Males, with smaller bodies, develop more quickly than females and fledge a few days earlier. After they have left the nest, the young return for food deliveries from their parents and to roost at night for at least ten more days. The siblings may remain together near the nest for five to six weeks, even after their parents have moved on. First-year birds tend to migrate earliest, and second-year birds migrate earlier than older ones.

Like Sharp-shinned Hawks, Cooper's Hawks hunt as they migrate, so a relatively high percentage of the ones passing by at Hawk Ridge are caught at the banding station. It's especially exciting on big migration days when the banders send a male and female of both species to the main overlook for release, allowing visitors to see the huge differences in sizes among the four. Some people believe that there is overlap between the largest female sharpies and the smallest male Cooper's, but comparisons in the hand make it clear that this is untrue. Then, one by one, those birds in the hand are released and take to the sky, and for one brief, shining moment we watch them in flight before their lives become truly their own again and they disappear into their future.

Cooper's Hawks at Hawk Ridge

Seasonal average sightings (over twenty years): 150

Earliest date of sighting: August 15

Latest date of sighting: November 21

Peak migration: September to mid-October

Record daily count: 44 on September 15, 1999

Record seasonal count: 319 in 1992

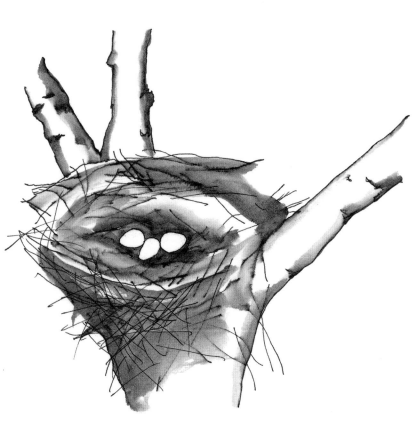

Northern Goshawk

Northern Goshawk (*Accipiter gentilis*)

For hawk aficionados, perhaps no sight is as thrilling as a Northern Goshawk in adult plumage winging past, its solid, steel-gray back and upper wings augmenting the regal authority in its slow, powerful wing beats and direct, purposeful flight. The white breast is so finely lined with black that we have the sense that the bird is garbed in a knight's chain mail. The bird's black helmet-like cap and wide black eye line lend it a forbidding aspect enhanced by the intense blood-red eyes.

The word *goshawk* is derived from the Old English name for the bird, meaning "goose hawk" or, possibly, "grouse hawk." It is pronounced gos-hawk, though people often say "Gosh!" when they see one. Goshawks are big and powerful enough to hunt grouse, ducks, small geese, rabbits, and hares. Medieval falconers of Europe treasured them both for their focus on large prey for the kitchen and for their courage and ferocity in the field. Goshawks often pursue birds in the air but will drop into the underbrush to continue a high-speed chase on foot. Attila the Hun's well-earned reputation for ruthlessness was enhanced by the goshawk adorning his helmet.

The ferocity of goshawks isn't limited to hunting. They are wildly protective of their nests, as bird banders and anyone who has walked near a nest can attest.

Fortunately for hapless people, they tend to nest in wild areas away from human intrusion. But unfortunately for them, disturbances from human activity, even as innocuous-seeming as camping nearby, can stress adults to the point that even after driving the humans away, the birds abandon the nest altogether.

Goshawks usually build their nest in the largest tree on their territory; the tree may be either deciduous or coniferous. They often repair a nest built a year or two earlier. Females do most of the construction, but males occasionally help. Throughout the time nestlings are in the nest, females bring live conifer sprigs to the nest, as several other hawk species do. The reason is not clear but may be related to controlling parasites or insect pests.

Females do most of the incubation, though males sometimes help when the female leaves. When the young hatch, the female broods them almost continually for at least nine days; if she does leave, the male occasionally broods them. The male provides almost all the food for the female during incubation, and for her and the chicks for the entire time they are small, and sometimes until the young fledge.

The young start taking short flights by the time they are a month old, and remain near the nest for another month before dispersing. The larger, slower-to-develop females remain on the territory longer. Their parents continue to feed them throughout much of this period. When young leave the nesting area earlier than normal, it is usually due to a food shortage.

Some Northern Goshawks appear to retain the same mate over many years, while others may change mates annually. Outside of the nesting period, adults are rather solitary and, like most raptors, don't remain with their mate through migration or winter.

A great many mysteries are associated with goshawk migration. Not all individuals migrate, even in large "invasion years." The exceptionally large migration in Duluth is thought to involve goshawks from western and central Canada skirting the Great Plains by moving east-southeast until they reach Lake Superior. Data supporting this include four goshawks banded at Hawk Ridge that were recovered in the northeastern corner of British Columbia.

Goshawk populations fluctuate wildly in approximately ten-year cycles that seem to coincide with the cycles of snowshoe hare and grouse, favorite prey species. Any cyclical population by definition has its ups and downs. But with goshawks, both the cyclic peaks and troughs are much lower than in previous decades.

In 1991 the U.S. Fish and Wildlife Service designated the Northern Goshawk a "Category II" species of concern, meaning additional information was needed before it could be listed as endangered or threatened, but in 1996 that designation was eliminated, and goshawks receive no special protection under the Endangered Species Act despite their precarious state.

We think of them as late-season migrants that pass through most abundantly in October and November, but occasionally we see one, usually a first-year bird, as early as August. Whether they are inexperienced birds just adjusting to independence from their parents or regal adults, every sighting of these massive birds inspires genuine awe of a kind reserved for those who are truly high and mighty. ◎◎

Northern Goshawks at Hawk Ridge

Seasonal average of sightings (over twenty years): 699

Earliest date of sighting: August 15

Latest date of sighting: December 11

Peak migration: October and November

Record daily count: 1,229 on October 15, 1982

Record seasonal count: 4,963 in 1972

Buteos

(Order Accipitriformes, Family Accipitridae)

The buteos, characterized by long, broad wings and a short, broad tail, compose the largest group of raptors counted at Hawk Ridge. This group includes several commonly seen species, and the annual totals of a single species, the Broad-winged Hawk, are usually much higher than the annual numbers of any other species at the ridge. But buteo migration tends to be concentrated on fewer days than that of other hawks. The large wings and wide tail of buteos combined with their relatively light weight give them exceptionally light "wing-loading," making them well adapted for rising on thermals and updrafts to cover long distances with minimal flapping. So their migration is heaviest on days with ideal weather conditions for thermal development along with westerly winds, which direct larger numbers of them to Lake Superior's North Shore.

Such large wings make buteos less maneuverable than most raptors. Whether searching for prey from a perch or from the air, when a buteo spots a likely target, it often folds its wings to drop like a bullet on it rather than flapping in pursuit.

Buteo plumage changes between a bird's first and second year, which confuses many hawk watchers. For example, not one of the Red-tailed Hawks hatched in a given year will have a red tail when it flies over Hawk Ridge in fall. To compound the confusion, many buteos display "color morphs"; that is, two or three entirely different color patterns may characterize different populations or individuals of a single species. And many individual buteos produce more or less than the normal amount of feather pigments, making them much darker ("melanistic") or lighter ("leucistic") than normal. Fortunately for hawk watchers, most buteos seen at Hawk Ridge wear the most characteristic plumage for their age and species, but the smattering of outliers provides exciting challenges, too.

Six species of buteos have been recorded at Hawk Ridge. The most abundant, the Broad-winged Hawk, may average close to fifty thousand a season, but its migration is extremely concentrated in mid-September.

Even during a week when tens of thousands are seen, there may be none at all spotted on days with rain or easterly winds. The other very common buteo is the Red-tailed Hawk; over eight thousand are counted each season, their migration spread out from early August through December. The Rough-legged Hawk is seen in much smaller numbers, averaging less than five hundred each season. Its migration is regular only in October and November.

The rarest buteo at Hawk Ridge, the Ferruginous Hawk, has never been documented from the main overlook, but twice, in 1975 and 1984, one was seen from the banding station, though neither was caught nor banded for verification. Red-shouldered and Swainson's Hawks are rare but regular; on average, three red-shoulders and six Swainson's are counted from the main overlook each autumn.

Broad-winged Hawk *(Buteo platypterus)*

On a mid-September morning when the sky is clear or partly cloudy and the wind is light and variable or westerly, wise hawk watchers hustle straight to Hawk Ridge. If the timing and weather are exactly right, they will witness one of the most spectacular phenomena in the natural world: Broad-winged Hawk migration.

The Broad-winged Hawk is by far the most abundant raptor at Hawk Ridge, providing both the highest annual numbers and the highest daily count of the season. The highest daily count is often well over a third of the seasonal total, and sometimes even 90 percent of it, which means Broad-winged Hawk migration may be concentrated into a few huge days each season. For hawk watchers during fall migration, this is the quintessential boom-or-bust species.

Most years, the biggest single-day broad-wing count will number from six thousand to ten thousand. In some years, a migration of several thousand may be repeated a few times, but in years when the highest daily count is over forty thousand, there may be only two or three other days with large movements.

On a good broad-wing day, birds often start moving by eight o'clock. The numbers don't start out big, but the first birds fly low in the sky, easy to observe closely. As the sun gets higher, its rays increasingly heating up the ground, thermals can grow strong enough to carry Broad-winged Hawks higher than the naked eye can see. By then, the closest broad-wings may look like pepper specks. On these days, an early-morning hourly count of a dozen birds may look more impressive than the later hourly counts that may reach ten thousand.

During September, the vast majority of hawks in any large kettle are Broad-winged Hawks. Living up to their name, they have long, broad wings, which are shown to perfection while floating on a thermal. From beneath, the wings appear pale, outlined in dark—almost as if a child were outlining a drawing with a black crayon. Adults have very broad, clean-looking black-and-white bands on their tails; birds hatched during that year (called first-year birds) don't have that conspicuous banding.

Ornithologists once believed that the Broad-winged Hawk was closely related to the Red-shouldered Hawk. The two species share a black-banded tail, and both feed heavily on cold-blooded vertebrates in wet forests. But chromosome analyses suggest that the Broad-winged Hawk is genetically distinct from all other buteos and is most closely related to Swainson's Hawk.

In big kettles, where the vast majority of birds are broad-wings, any other buteos will be noticeably bigger, and accipiters and falcons will be noticeably different in shape. On busy days, one of the official counters will be counting Broad-winged Hawks; one will be tallying sharp-shins, which are almost always numerous in September; and another one will be picking through the birds to identify and count all the others.

Broad-wings migrate en masse because they must travel all the way to South America, an exhausting distance if they were to power their flight with laborious wing beats. Instead, they migrate only under ideal conditions, when they can glide to a thermal, rise to gain as much altitude as possible, glide to the next thermal, and on and on, easily covering a hundred miles or more in a single day.

Why do they migrate such long distances? Broad-winged Hawks feed heavily on cold-blooded invertebrates, especially frogs, toads, and snakes, as well as rodents, large insects, young birds, and other relatively small prey. Cold-blooded vertebrates can be very hard to find on the forest floor even as far south as Florida in winter, so the mass of broad-wings heads to the tropics. Migration counts over Veracruz, Mexico, and over the Panama Canal are much larger than the best days at Hawk Ridge.

When they return in spring, broad-wings set up housekeeping on their territories fairly quickly. This nesting hawk common in moist woodlands of northern Minnesota and Wisconsin can be difficult to see on its breeding territory, but occasionally one sits on a power pole along a forested road or highway, or on a branch above a path within a wooded area. Broad-wings are fairly calm and often sit tight for close viewing. Their call is a piercing, down-slurred whistle that doesn't sound hawk-like.

Some pairs of Broad-winged Hawks mate together for at least a few years, but others apparently select new mates at the start of each season. During the nesting season, pairs are fairly solitary, with few or no interactions with birds on neighboring territories. Both the male and female construct the stick nest in thick branches of a conifer or deciduous tree, though the female tends to do the most work. As the nest nears completion, the female carries in fresh conifer sprigs, which aren't built into the nest but, rather, rest on the nest rim. Only females have a brood patch, an area of featherless skin on their underside that transfers heat to eggs and young, and they do most of the incubation, though males sometimes incubate while females are off the nest. Males may finish incubating if a female dies, but that probably merely delays the inevitable loss of the chicks, because males don't typically feed them. Females tear prey into appropriately sized chunks for the first two or three weeks and then allow the chicks to tear up prey animals for themselves. Small mammals and toads are the most common prey brought to chicks.

Broad-wing chicks are fairly helpless on hatching, but within the first day they can bite hard. In most nests, two to three chicks hatch. In about 10 percent of nests, chicks fight aggressively, and in some cases the smallest may be killed. This is most likely to happen when food is scarce.

Chicks first leave the nest at about four weeks of age, are good at flying by six weeks, and start flying farther from the nest tree by seven weeks, when they start capturing their own prey. They remain on their parents' territory for another week or so. As August draws to a close, they begin wandering. Migration is so concentrated that on big days it's impossible to keep track of how many are first-years and how many are adults; there is no way to distinguish the sexes in flight, so we don't know if there is any difference between the departure timing of any groups. Each one is headed thousands of miles away, where little is yet known about broad-wing winter behavior and ecology. It's intriguing that so abundant a species can still keep so much of its life shrouded in mystery.

Broad-winged Hawks at Hawk Ridge

Seasonal average of sightings
(over twenty years): 48,081

Earliest date of sighting: August 1

Latest date of sighting: November 1

Peak migration: September

Record daily count: 101,716 on September 15, 2003

Record seasonal count: 160,703 in 2003

Red-tailed Hawk *(Buteo jamaicensis)*

In Duluth, winter garb comes out in early fall. Long johns, winter jackets, and thick hats and gloves are called for on blustery October days. Yet even on the coldest days, dozens of people still gravitate to Hawk Ridge to stand in the biting wind simply to witness Red-tailed Hawks passing through. Sometimes thousands fly over in a single day.

The red-tail is the hawk "making lazy circles in the sky" in the song "Oklahoma!" This large buteo is striking even when the lighting or angle obscures the red tail, but when an adult banks, showing that tail to perfection, even old-timers who have seen thousands may feel their heart swell.

Red-tailed Hawks still in their first year lack the red tail. At any age, most of them can be recognized by the delicate markings across the lower front, producing what looks like a belly band. All red-tails also have dark markings on the leading edge of the wing, from just past the neck to the bend of the wing. This "patagial mark" is a useful characteristic for identification. Of course, the red tail is a perfect giveaway. The only other raptor with a bright rusty red tail is the tiny American Kestrel. Some people new at bird identification mistake the kestrel for a "baby red-tail," but by the time red-tails fledge from the nest, they are fully as large as their parents.

Of all the raptors of North America, the Red-tailed Hawk is probably the one most frequently encountered by people, adapted to surviving almost everywhere, from wilderness to major urban centers. This is the large hawk seen along highways just about anywhere in America.

Red-tailed Hawks hold an almost magical fascination for people. One famous red-tail nicknamed Pale Male established residency on a New York City building on Fifth Avenue across from Central Park in 1991; still living in 2012, he has been the subject of books and at least one documentary film. A video camera set up by the Cornell Lab of Ornithology to provide Internet streaming of nesting activities by a pair of Red-tailed Hawks on the Cornell University campus attracted over 172,000 unique viewers from 143 countries during just the first week after it was launched in 2012, before it was even publicized.

Red-tailed Hawks are conspicuous enough in many places that even nonbirders can thrill at their spectacular aerial displays during courtship, notice their large stick nests in large trees off roadways, and watch individuals swoop down to grab prey in weedy interstate

highway medians as cars race by. Adaptability is the watchword for this magnificent raptor.

As famous as red-tails are for circling in the sky and as acute is their vision, they seldom hunt from such heights—their large wings are too unwieldy for rapid maneuvering. They usually perch in tall trees or on power lines to watch for prey. Red-tails hunt small and medium-sized animals, including rodents, rabbits and hares, birds, reptiles, and amphibians. They can manage prey as large as pheasants and jackrabbits. Some have adapted to eating fresh carrion, especially when they see an animal struck by an automobile or are alerted to a carcass by magpies, ravens, and crows. Red-tails usually carry small prey to a feeding perch. They swallow small mammals whole but behead and pluck birds before eating them. Larger prey is at least partly eaten on the ground, but they may carry chunks to a perch to eat.

The life of a predator is dangerous. One red-tail was found dead with a coral snake in its talons, at least one was killed when it flew full force into a porcupine, and there are reports of some killed after attacking rattlesnakes. In one case, a healthy-appearing red-tail was found dead; necropsy revealed that it had apparently swallowed a still-alive pack rat, and the rodent clawed and bit through the hawk's esophagus.

As perilous as the life of a hunter is, once a red-tail has survived a year or two, working out strategies for finding food and avoiding danger in all seasons, its life expectancy goes up dramatically. The famous Pale Male still nesting on Fifth Avenue was at least twenty-two years of age in 2012. Several banded red-tails have survived well into their twenties, one over thirty years.

Thousands of Red-tailed Hawks pass over Hawk Ridge each fall, but the species is considered a short- to intermediate-distance migrant, and individuals from many areas, including northern Minnesota, are nonmigratory. Many red-tail pairs remain together for life, maintaining their bond year-round. Their aerial courtship displays are most common in spring but may be seen at any time of year. Red-tails are early nesters, starting breeding in February or March in many places, and some Wisconsin pairs have been documented breeding as early as late January. They raise one brood a year, but pairs may build more than one nest in a single year—long-established pairs may have several old nests to choose from. They may add new sticks and fresh greenery in more than one of their nests before the female finally selects one in which to lay eggs. In Minnesota, red-tails normally choose the tallest trees in their territory for nests, but where trees are scarce, they may nest on cliffs, transmission line towers, and man-made structures projecting above the landscape. Wherever they nest, red-tails seem to want unobstructed access to the nest from above and a broad view of the landscape.

Both the male and female build or refurbish the nest, the female doing most of the work in the bowl. Like several other hawks, they set fresh conifer sprigs along the nest rim, but no one understands the function of these. Some speculate that the greenery may repel some parasites, others that it may alert potential nest competitors that the nest is taken.

Red-tails typically raise two or three chicks, and rarely four. Both parents have a brood patch, and both incubate the eggs, beginning when the first is laid. Incubation lasts about a month. The male often provides the female with food during this time, and sometimes she brings food to him. After the chicks hatch, the female broods the young, while the male provides food for the family. For the first four or five weeks, she tears prey apart for the chicks. After that, the food is deposited in the nest, and the chicks tear it apart. When food is scarce, the youngest chick often dies.

Young red-tails generally leave the nest when about six weeks old but stay very near for a few days. Parents usually drop food near the fledglings rather than feeding them or giving it directly to them. About six or seven weeks after fledging, the young begin hunting, though their parents continue to provide the bulk of their food for another week or two. By summer's end, these young birds are ready to take to the skies. Young red-tails often disperse from the breeding area and migrate ahead of their parents.

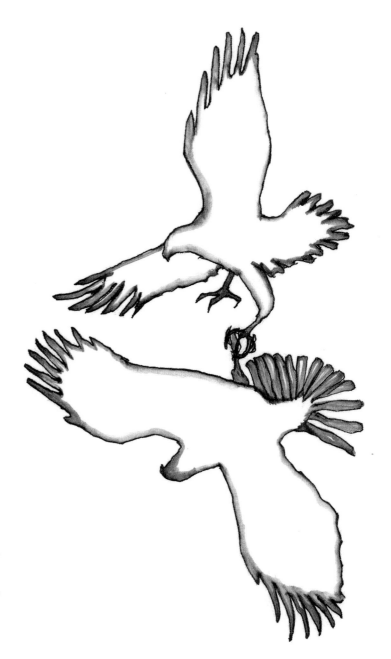

Because they are so adaptable to the kinds of changes people make on the landscape, Red-tailed Hawks have displaced Red-shouldered, Swainson's, and Ferruginous Hawks where natural habitat has been altered for urbanization and agriculture. Their population is considered secure, comforting news for all who thrill at the passage of this magnificent bird through the Hawk Ridge skies.

Red-tailed Hawks at Hawk Ridge
Seasonal average of sightings (over twenty years): 8,254
Earliest date of sighting: August 1
Latest date of sighting: December 12
Peak migration: October
Record daily count: 3,988 on October 24, 1994
Record seasonal count: 15,358 in 1994

Rough-legged Hawk (Buteo lagopus)
The arctic tundra is a forbidding landscape even in summer; birds that breed there must be adapted to wet, cold conditions even in July. The hawk most perfectly adapted for raising chicks in those conditions, heavily insulated with thick feathers all the way down to its toes, is the Rough-legged Hawk. Nicknamed the arctic fluff ball, this hardy species is named for the dense feathers on its legs, though the word *rough* hardly seems an apt description for such puffy, soft plumage.

Long winter nights make wintering in the Arctic Circle too forbidding for Rough-legged Hawks. This species is a "complete migrant": the entire population leaves the breeding ground for a separate wintering range spanning virtually all of the western and central United States and the northern two-thirds of the East up to and just beyond the Canadian border. During some winters, large numbers of rough-legs remain in Minnesota and are easy to find in open fields, pastures, and grasslands; in other years they are few and far between. The cyclic nature of their migration and wintering patterns appears to be influenced by the cyclic populations of their preferred prey species, lemmings and voles.

Rough-legged Hawks have two common color "morphs." From beneath, the less common dark morph's belly and wing linings are fairly solid blackish brown, with paler flight feathers tipped with dark, and its tail can appear whitish with a black terminal band. The more common light morph has a white tail ending in a dark terminal band, a blackish belly, and a distinctive large, dark "carpal patch" in the center front of each wing; the outer tips and trailing edge seem outlined in

black. Rough-legs often glide with their wings slightly uptilted. They can also hover in place for many seconds or even a full minute while flapping their wings.

Rough-legged Hawks have relatively heavy bodies for the size of their wings, so they rely more on flapping than floating on thermals, though they do sometimes kettle at Hawk Ridge. Because they are less dependent on thermals than most hawks, they often are seen migrating earlier and later in the day than other buteos.

Rough-legged Hawks nest in exposed areas near the tops of cliffs and outcrops on the tundra. The male selects the nest site, and the pair works together on construction. The male brings a variety of building materials including sticks, caribou bones, and soft materials for lining, and the female fashions them into a bulky mass. The first eggs are laid sometime in May, depending on when the snow melts. An average clutch size is three, but in some years rough-legs may produce as many as seven eggs or as few as one, probably related to availability of their preferred prey, lemmings. They also eat voles, arctic ground squirrels, and a variety of birds, including ptarmigans.

Females do virtually all the incubating, leaving the nest only to eat food brought by the male and occasionally to gather additional nesting materials. After the chicks hatch, she broods them almost all the time. The male does most of the hunting, and both male and female tear apart prey, mostly lemmings, to feed the young. By about two weeks of age, the chicks can swallow hamster-sized lemmings whole.

In about six weeks, the chicks can fly but usually stay close to the nest for another week and remain with their parents a few weeks more, sometimes through the start of migration. By the time it reaches Hawk Ridge, a Rough-legged Hawk has already flown a thousand miles or more over Canada. Some will remain in Minnesota through the winter; others will move a thousand miles beyond, their wing beats stitching the seams connecting nations, provinces, and states into one natural world.

Rough-legged Hawks at Hawk Ridge

Seasonal average of sightings: 497

Earliest date of sighting: August 30

Latest date of sighting: December 13

Peak migration: October and November

Record daily count: 204 on November 10, 1963

Record seasonal count: 1,026 in 1994

Red-shouldered Hawk (*Buteo lineatus*)

The Red-shouldered Hawk is one of the more abundant raptors in the eastern United States and southeastern Canada, but seeing one at Hawk Ridge is a matter of luck. The main population of Red-shouldered Hawks extends east and south from the Mississippi River beginning at about the Twin Cities through the Great Lakes to Nova Scotia. Most of the species is nonmigratory, but those in central Wisconsin through New England and down through the Appalachians do head south for winter. Hundreds pass over Hawk Mountain in Pennsylvania annually, but Hawk Ridge is off their course. Prevailing westerly winds keep virtually all of them well east and south of Duluth, but on average three individuals wander by the ridge each year. The species has a curiously disjunct distribution: although the main population is in the East, an isolated, mostly nonmigratory population is also found in the far West, mostly along the Pacific Slope.

The ideal habitat for Red-shouldered Hawks is bottomland hardwood forest, flooded deciduous swamp, and other wooded areas along rivers and streams. The Florida subspecies is a common breeding hawk in parks, neighborhoods, and golf courses. These hawks are generalists, taking a variety of small animals but most abundantly rodents, frogs, and snakes. In some areas, they focus seasonally on birds, crayfish, or insects. Like most forest buteos, they usually hunt from a perch, scrutinizing the ground until they spot prey, then swooping down to grab it.

When a Red-shouldered Hawk does approach Hawk Ridge, one of the counter's assistants or the naturalist virtually always shouts it out. Hawk watchers all focus on the large buteo with rusty belly and wing linings and striking horizontal black-and-white tail stripes. They savor the sight and when they leave for the day, feel as if they have won the lottery.

Red-shouldered Hawks at Hawk Ridge

Seasonal average of sightings (over twenty years): 3

Earliest date of sighting: August 25

Latest date of sighting: November 10

Peak migration: October

Swainson's Hawk *(Buteo swainsoni)*

Every now and then on a September day dominated by west winds, someone will shout out an exultant "Swainson's Hawk!" If the bird is indeed a Swainson's Hawk, the hawk watchers' joy will be palpable.

Swainson's Hawks breed throughout the Great Plains and beyond, in open grasslands and sparse scrublands of the American West. They eat ground squirrels, pocket gophers, and other small animals, including many that are considered agricultural pests. They winter primarily on the pampas of Argentina, where they feed almost exclusively on insects, especially grasshoppers.

Of all the hawks that pass over Hawk Ridge, this bird is probably the most helpful to humans in terms of economic interests, but ironically it is also the one suffering the most from human activities. Until the late 1930s, people shot Swainson's Hawks as "varmints." Now direct persecution has become relatively rare, but these beautiful hawks are still injured or killed in collisions with cars and trains, electrocuted by power lines, and caught on fences.

More significantly, Swainson's Hawks have been inadvertently killed in huge numbers by pesticides on their wintering grounds. In 1995 and 1996, organophosphate insecticides used to control grasshopper outbreaks killed thousands of Swainson's Hawks in Argentina. Some deaths resulted immediately when foraging hawks were sprayed by pesticide applicators. Other deaths occurred within days, when the hawks ate poisoned grasshoppers. Nearly six thousand carcasses were found, and it is believed that an estimated twenty thousand were killed by pesticide applications in those two years. Overall, the species has been declining steadily.

Swainson's Hawks tend to migrate straight to Texas and beyond from their western range; Hawk Ridge is well off course for them. Sightings at the ridge are weather dependent. In 1988, an amazing thirty-three flew past, thirty-two in a single day, due to an exceptionally long period of sustained westerly winds. As they grow rarer, seeing them grows less likely each year. But any time one of these beautiful hawks wings past, attention will be paid. ◎◎

Swainson's Hawks at Hawk Ridge

Seasonal average of sightings (over twenty years): 6
Peak migration: September
Record seasonal count: 33 in 1988

Falcons

(Order Falconiformes, Family Falconidae)

ong, narrow, pointed wings and a fairly long, narrow tail uniquely adapt falcons for speed. Ornithologists traditionally placed all the raptors (except, for a time, vultures) in a single order, the Falconiformes. But recent DNA studies have conclusively determined that falcons are not related to the others. Some may find it surprising that falcons are genetically much more closely allied with parrots and songbirds than with hawks. In 2010 the American Ornithologists' Union created a whole new order, the Accipitriformes, in which they placed all the nonfalcon hawks.

During their first year, falcons of most species can be distinguished from adults. One species, the American Kestrel, shows strong sexual dimorphism; males are much more boldly patterned and brightly colored than females.

Hawk Ridge's two tiniest falcons, the American Kestrel and Merlin, are often seen capturing dragonflies as they cruise past the main overlook. They don't bother to find a perch to eat such tiny prey—they simply pull their talons up and put their head down momentarily, nibbling away as they continue on. Ideal days for seeing both species often coincide with major dragonfly flights.

Peregrine Falcons were once exceptionally rare at Hawk Ridge, though they were still seen yearly even after the species had been extirpated from most of North America; the peregrines flying over Hawk Ridge were migrating from tundra breeding grounds. Thanks to Peregrine Falcon reintroduction projects, the species is now much more common, with some seasonal totals exceeding a hundred. In 1996, the Minnesota status of the peregrine was changed from endangered to threatened. In 1999, the Peregrine Falcon was removed from the federal endangered species list.

Two extremely rare falcons have been reported from Hawk Ridge. Gyrfalcons, from the far North, have been seen eleven times since 1981; Prairie Falcons, eight times since 1979.

American Kestrel *(Falco sparuerius)*

When someone calls out "kestrel!" at Hawk Ridge, everyone's binoculars gravitate to the bird as if it were a rare species. American Kestrels are one of the common migrants at Hawk Ridge, but regardless of how many pass through on a good day, their lovely colors and dainty flight make them guaranteed crowd pleasers. And hawk watchers can't help but smile to see one of these exquisite birds snatch a dragonfly out of the sky. The bird flies along with its tiny prey in its talons, every now and then drawing its legs forward and thrusting its head down to grab a bite.

Huge swarms of dragonflies moving along the Lake Superior shoreline in August and September are a vivid reminder that birds are not the only migratory animals. The common green darner is the most well-known dragonfly that passes along Hawk Ridge in advance of cold fronts in August and September, and we humans are not the only ones who take notice. From the time Hawk Ridge started attracting birdwatchers, people have been noticing kestrels munching on dragonflies. In 1995, official counter and Hawk Ridge bird bander Frank Nicoletti began a study carefully counting dragonflies as well as raptors passing by. He found that American Kestrel migration peak days coincide with peak days for green darner migration. Closely studying their habits, Nicoletti found that at midday, when migration conditions are ideal, kestrels don't eat many

dragonflies; they are focused on migrating, flying higher in the sky than they do later in the afternoon. But when they come lower, Nicoletti noticed, a large number of them feed on dragonflies as they move along.

The word *kestrel* is probably derived from the French word for a leper's "clicket," a clapper once used to warn people of a passing leper; the name was originally given to the European Kestrel for its noisy call. Whether in Europe or America, kestrels attract our notice. "The prettiest and jauntiest of our hawks, and yet no prig," is how ornithologist Elliott Coues described them in 1874. Naturalist William Brewster, in 1925, called it "most light-hearted and frolicsome." Outside of migration, we see them along roads and highways, perched on wires or snags, or hunting over farm fields, pastures, and other open habitats, often hovering in place for many seconds at a time. This dainty predator feeds primarily on insects whenever they're available. Kestrels also feed on many small rodents and songbirds, especially in winter.

When perched on wires in the characteristic falcon pose described by ornithologist Winsor Marrett Tyler as "hunched up and frowning," the red tail found in all kestrel plumages can be conspicuous. People who don't know how young birds develop often think they are seeing a "baby Red-tailed Hawk" when they spot a kestrel. Red-tails are fully adult size when they leave the nest, and don't have the distinctive red tail until they are adults. Kestrels are an entirely different species despite the two sharing the trait of rusty tails and being the two raptors most often seen along roadsides.

Kestrels are cavity nesters, using natural cavities in trees, woodpecker holes, and nest boxes to raise young. They nest in open country, and their populations did much better during the decades when farmers set aside uncultivated fencerows and used minimal pesticides. Now the species has been declining across the range, and Hawk Ridge numbers are slowly decreasing. People in appropriate habitat can help local populations by setting out kestrel boxes for them. Carrol Henderson's *Woodworking for Wildlife,* published by the Minnesota Department of Natural Resources, has both building plans and suggestions for placing boxes to make them most attractive to and useful for the birds. The bottom of the box should be lined with wood chips, straw, or other soft materials.

Female kestrels are larger than males and about 10 percent heavier. As long as both birds in a pair remain alive and return to the same nesting area in spring, their pair bond seems to continue from year to year. A few pairs also apparently remain together during winter, but they are the exception. Like many birds of prey, most kestrels maintain an individual winter territory,

and females can be extremely aggressive defending it. The Lake Superior Zoo's two were quite content in adjacent enclosures, but when put together in a very large one, the female continuously harassed the male until the two were separated again.

Kestrels have more predators than most raptors have to contend with. Large snakes sometimes devour eggs or chicks, and fire ants invade nests in southern areas. Full-grown American Kestrels have been found in the stomach contents of Northern Goshawk; Red-tailed, Sharp-shinned, and Cooper's Hawks; Peregrine Falcon; and Barn Owl. One American Crow was documented killing and eating a female kestrel.

American Kestrels are fully protected, but some shooting still occurs, and this may be exacerbated in states with Mourning Dove hunting seasons—the two species are often confused thanks to their long tail, pointed wings, and habit of perching on wires. Kestrels are also exceptionally vulnerable to pesticides, partly because when hunting birds and mice, they are especially drawn to those moving slowly and acting unusually; so some studies have shown that they feed preferentially on poisoned prey. Thanks to withdrawal of some of the most dangerous pesticides, the main cause of the kestrel's slow population decline seems to be habitat degradation and loss. Fortunately, the species is still considered reasonably secure.

In spring, each male kestrel scrutinizes his territory to locate and inspect potential nest cavities. When his established mate arrives or after he attracts a new one, he escorts her to each cavity, and she makes the final decision about which will be used. American Kestrels do not bring nesting materials to the cavity, using any loose materials on the floor to make a shallow depression. Kestrels do occasionally lay eggs on the bare floor of an empty nest box, but these have low hatchability, probably due to chilling or breakage.

A normal clutch size is four or five eggs, usually laid every other day. Both the male and the female have incubation patches, and both incubate, although the female does so more than the male. In one study, males were found to incubate more on sunny than overcast days. After the chicks hatch, the female broods for a week or longer, while the male is the sole provider for the family. Both parents feed the chicks. When food is scarce, the last-to-hatch chick often dies and may be eaten by larger siblings.

Chicks fledge at about a month of age, continuing to depend on their parents for feedings for another two weeks as they start to hunt on their own. They test out their skills on various moving objects and are often seen pouncing on butterflies and other insects. During this time, the siblings usually roost close together. They grow less sociable as they become more adept at hunting. In fall, some young kestrels associate together in small hunting groups, but these birds are usually not related to one another.

At Hawk Ridge, kestrels sometimes dive-bomb the owl decoy, giving visitors a visual treat. These little falcons are often caught at the banding station, and when one is brought to the overlook for release, the close views, especially of colorful adult males, elicit exclamations. Like virtually all birds, kestrels appear far tinier in the hand than when they open their wings and take to the sky. This is mostly an optical illusion, but the nearly two-foot wingspan is impressive on this ten-inch bird. Whether they are cruising past on a mission or fluttering leisurely along munching dragonflies, American Kestrels are a delightful splash of color on the Hawk Ridge skyline.

American Kestrels at Hawk Ridge

Seasonal average of sightings (over twenty years): 1,909
Earliest date of sighting: August 1
Latest date of sighting: November 11
Peak migration: September
Record daily count: 744 on September 9, 2002
Record seasonal count: 3,637 in 2002

Merlin *(Falco columbarius)*

A diminutive falcon approaches the main overlook. Poor lighting makes it impossible to pick out color or even to discern the absence or presence of a tiny row of white dots along the trailing edge of the wings. The dots are characteristic of American Kestrels, and their clear absence combined with overall dark plumage would indicate that the bird is a Merlin. Even without those important details, experienced hawk watchers recognize it almost instantly by its powerful, rapid wing strokes. Hawk authority Pete Dunne describes the difference in flight between a Merlin and an American Kestrel as the difference between a Harley-Davidson and a scooter. This bird is no scooter.

Merlins are our only woodland falcon, their rapid, powerful flight perfect for pursuing prey through dense forest. They are only an inch or so longer than kestrels but outweigh them by almost 40 percent, and the difference seems to be all muscle. Once called pigeon hawks, Merlins are quite capable of bringing down Mourning Doves weighing 70 percent of the Merlin's weight, and occasionally even heavier Rock Pigeons.

The name *Merlin* is not at all etymologically related to King Arthur's wizard's name, but to the Old French *esmerillon,* a word coined for the species many centuries ago. Despite its tiny size, the Merlin has long been popular among falconers, particularly as a "lady's hawk"; Catherine the Great and Mary Queen of Scots were famous enthusiasts. Most modern falconers prefer larger species, but some still appreciate the Merlin's speed and tremendous heart.

One population of Merlins has long been urban dwellers in Saskatoon, Saskatchewan, but the species as a whole was generally considered a bird of wild forests and prairies and to be rare even within those habitats. In 1932, Thomas S. Roberts wrote in *The Birds of Minnesota* that he had seen Merlins "not over a half a dozen times during fifty years." But since the mid-1980s, Merlins have been nesting in small towns and cities along Lake Superior. In Duluth, Merlins currently nest in virtually every neighborhood. One year, a nest was in a tree next to the bus stop serving a large junior high school on a busy road. Merlins are exceptionally noisy, their high-pitched calls sounding rather like a hyperactive Killdeer, but Duluth residents have become accustomed to them. Despite how the population has grown since the 1980s, based on numbers counted at Hawk Ridge, Merlin numbers are overall stabilizing and even declining.

Like other falcons, Merlins don't build their own nest. Instead, they use abandoned crow nests built the previous year, seeming to prefer those that are well concealed yet provide a good view of the surrounding area. They produce an average of four eggs in their only clutch of the year. Both sexes incubate, the females longer than the males each day. Merlins hunt for birds almost exclusively, though one was seen in the neighborhood under Hawk Ridge preying successfully on a red squirrel, and they are often seen eating dragonflies at Hawk Ridge.

During incubation, the male brings an average of three birds a day to his mate. Eggs hatch in about a month, and similar to many raptors, the male continues to do most of the hunting for the family while the larger female guards and broods the young; he brings an average of nine or ten birds a day to the nest while chicks are present. The young fledge when about a month old, remaining dependent on their parents for weeks longer. Adults spend the night roosting away from the nest, while the chicks roost together.

One Merlin nesting in the neighborhood below Hawk Ridge discovered an ideal method for hunting at a bird feeder in a side yard two blocks from its nest. It would speed down the road above the sidewalk, dropping so low that its wings almost seemed to brush the pavement, and suddenly rip around the corner of the house at such speed that birds at the feeder never saw it coming.

People sometimes feel irritated or saddened when a hawk or falcon preys on their backyard birds, which is understandable from the viewpoint of the prey, though Merlins might wonder what better function a bird feeder could serve than to feed birds to hungry birds. The lives of predators are treacherous. In one high-speed chase, a Merlin in Duluth became entangled in a macramé hammock, wrenching its wings; it survived only because of the quick action of two young girls. In Europe, a Merlin was once recorded stealing the catch from another hawk. The Merlin was then robbed by a Honey Buzzard, which was in turn robbed by a Peregrine Falcon—a surprising twist on the concept of a food chain.

At Hawk Ridge, Merlins are one of the raptors most likely to attack the owl decoy. Hawks quickly discern that the owl isn't a real predator, and after swooping in once or twice, most move on. But some Merlins have dive-bombed the owl more than twenty times—they seem to get even angrier after figuring out the decoy is a fake.

Merlins at Hawk Ridge

Seasonal average of sightings (over twenty years): 216

Earliest date of sighting: August 13

Latest date of sighting: December 3

Peak migration: September to mid-October

Record daily count: 73 on October 9, 1997

Record seasonal count: 362 in 1997

Peregrine Falcon *(Falco peregrinus)*

Of all the birds in the world, the one most famous for speed is the Peregrine Falcon. Visitors at Hawk Ridge are sometimes surprised to see a peregrine leisurely circling within a kettle of hawks or winging past at the species' standard flapping-flight speed of just twenty-five to thirty-five miles per hour. That is hardly moseying along but is nothing like the two hundred miles per hour speeds reported in books.

Peregrine Falcons feed on a wide assortment of prey, mostly birds. Although they have been documented killing and eating prey from tiny hummingbirds to huge Sandhill Cranes, the vast majority of their quarry is medium-sized birds, including pigeons and doves, waterfowl and shorebirds, songbirds, and American Kestrels.

Even during a hunt, Peregrine Falcons seldom fly two hundred miles per hour. When a peregrine locates potential prey, it may simply and directly pursue it at speeds of less than seventy miles per hour. Direct pursuit is a common technique used when the quarry is a relatively slow-flying bird, such as a cuckoo or jay. In other situations, the falcon may use a surprise attack, surreptitiously using concealing features of the terrain, including buildings, banks, and ridges, to remain hidden until so close that the prey doesn't know what hit it.

When chasing a flock of birds such as waxwings, shorebirds, or starlings, a peregrine may use a technique called "shepherding." Rather than risk an inadvertent collision with one bird while pursuing another within a dense flock, the peregrine repeatedly dives at the periphery; when a panicked bird breaks formation, the falcon may be able to grab it.

A hunting technique called "ringing up" is another tool in the peregrine's bag of tricks, used to pursue a bird that is flying higher than the falcon. The peregrine may circle up until it is slightly above the prey and then make repeated shallow dives on it until the target becomes exhausted and more easily caught. Or the falcon may start out directly pursuing the bird, especially where there is little ground cover, and if the prey tries to escape by rising up in tight spirals, the falcon will make wider rings around it, maneuvering to keep the prey bird in flight until it becomes exhausted.

Witnessing any of these hunting techniques is gripping, but they pale in comparison with seeing a peregrine rise above its quarry, suddenly fold its wings against its body, and drop in a high-speed stoop. Stoops at shallow angles are impressively fast, mostly at about 66 to 86 miles per hour. Perfectly vertical stoops seem to violate the rules of physics. Theoretically, the fastest

a peregrine-sized free-falling body could drop in vertical descent is about 238 miles per hour. In reality, some Peregrine Falcons have been clocked going faster than that—in one case, a whopping 273 miles per hour. In 1999, Ken Franklin, a free-fall parachutist, studied trained falcons to learn how they adjust their bodies to control speed. He found that at 150 miles per hour, a peregrine assumed a diamond shape, its wings tucked and shoulders slightly extended. Above about 200 miles per hour, it became hyperstreamlined, pulling its wings tight to its body and extending its head to assume a more elongated position.

Flying at such amazing speeds requires several adaptations for survival. For example, the changes in air pressure in a high-speed dive could damage the lungs if not for bony tubercles on the peregrine's nostrils.

They guide the powerful airflow away from the nostrils, enabling the bird to breathe more easily and reducing the change in air pressure. The peregrine's nictitating membrane is translucent to permit it to keep its eyes moist and debris free without compromising vision. And its advanced nervous system can process all the sensory cues necessary to adjust flight direction and speed even as it homes in on unpredictable prey.

How does a peregrine kill or seize prey at such high speeds without being injured? To grab quarry outright, the falcon must be flying close to the same speed, or its own momentum will cause the prey to pull out of its talons. The peregrine often flies in under its quarry and rolls over or flips up to grab it from below or the side. It may also grab it from behind, holding onto the wings or neck of larger birds. High-speed cinematography shows that all four toes are widely splayed at the moment of contact and then immediately formed into a fist after the blow.

Once prey is caught, the falcon bites into the neck using its "tomial tooth," a small notch and projection on either side of the bill just behind the tip. This tiny tool is effective at breaking the spine and base of the skull of fairly large prey and also at breaking long bones before swallowing.

Peregrines often eat bats on the wing. One male hunting over the Colorado River in the Grand Canyon caught and ate seven bats in twenty minutes in uninterrupted flight. Peregrines carry most of their avian prey to perches to eat. Small birds are usually plucked and then the head and pectoral muscles devoured. The gut is often pulled and discarded, but other muscles and bones may be eaten. Larger prey, such as the Sandhill Crane one observer documented, may be partly eaten on the ground before parts or all of the remainder is carried to a feeding perch or the nest.

Historically, Peregrine Falcons nested on cliffs along mountain ranges, river valleys, and coastlines, perhaps most extensively across the arctic tundra, northern Pacific coast, and along the Rocky Mountains. In 1936, Thomas S. Roberts wrote in *The Birds of Minnesota* that about six pairs nested along Minnesota's Lake Superior shoreline, probably about the same number in bluffs along the Mississippi River south of Red Wing, a few more pairs along the upper St. Croix River, and a small number in the cliff-bordered lakes and streams between the Canadian border and Lake Superior. Roberts concluded that these occupied nests were probably little changed from previous generations and speculated, "It is likely that successive generations [of Peregrine Falcons] will continue to occupy them in the future, as the aeries are well-nigh inaccessible and their owners rarely expose themselves to danger."

In 1936, Roberts could not have predicted the insidious threat to Peregrine Falcons looming on the horizon. Although DDT was first synthesized in 1874, its insecticidal properties weren't discovered until 1939, and it wasn't available for use in agriculture and for mosquito abatement until after World War II. But within three decades, the Peregrine Falcon had been entirely wiped out in Minnesota and the rest of the United States east of the Rocky Mountains.

If Roberts could never have predicted the extirpation of Peregrine Falcons from the state, neither could he have predicted their spectacular comeback. How could he have dreamed that a bird that "keeps largely to the wilds and is not numerous enough to attract much attention" would become a beloved resident in the state, now nesting successfully in more urban sites in Minnesota than had ever existed in the wild in the entire state before?

In 1936, the same year that the second edition of *The Birds of Minnesota* was published, a wild pair of peregrines nested on the Sun Life Building in Montreal—the first known incidence of fully wild individual peregrines nesting on a man-made structure anywhere in the world. The birds had been hanging around on the roof for a few weeks, so researchers at McGill University placed a rough box on the twentieth floor, on the side of the building facing Central Station, and filled the box with sand and gravel. The situation quickly became an example of "if you build it, they will come." The birds nested on the Sun Life Building every year through 1952. During that time they laid about fifty eggs, of which about half hatched. Many of their egg failures may have been due to the increasing levels of DDT during that time, and when they disappeared, no falcons replaced them.

Thanks to that one pair of falcons, when the first Peregrine Falcon reintroduction projects were launched, scientists realized that they would not be restricted entirely to releasing and providing food to young chicks (in a process called "hacking out") on difficult-to-reach cliff sites. From the start, many of the young falcons produced by falconers' birds were released in urban areas. The projects begun in the late 1970s and throughout the 1980s were so successful that a great many cities in the United States and Canada now have nesting peregrines. In Minnesota, one of the natural sites where young falcons were introduced was on Palisade Head up the shore from Duluth. Little by little, a population in the state became established.

Now a few pairs of peregrines nest in Duluth. The Raptor Resource Project constructed a nest box downtown on the Greysolon Building in the late 1990s to see if peregrine offspring from birds in the reintroduction project might move in, and in 2003, a pair of falcons did. The first two birds to use the nest were unbanded, their origins unknown. Hawk Ridge Bird Observatory, in a collaborative effort with the City of Duluth, has been providing educational programs and informal interpretation in downtown Duluth; naturalists with spotting scopes and binoculars point out the movements and activities of Duluth's most visible raptor family. During early autumn, some of these birds occasionally fly past Hawk Ridge.

Peregrines that breed on the tundra are the longest-distance migrants of the species, inspiring the species' name—*peregrine* means "wanderer." Some of them winter in South America. Many of the peregrines that have been reintroduced belong to subspecies that are less strongly migratory, and some of these winter in Minnesota. There are usually one or two peregrines in the Duluth harbor throughout winter.

Visitors at Hawk Ridge can seldom see bands on raptors as they fly through—the legs are obscured by belly feathers. Whether a passing bird is one of "ours" from Minnesota or a wanderer from the tundra, every peregrine winging past is living proof that when a critically endangered species gets protection under the Endangered Species Act, miracles do happen.

Peregrine Falcons at Hawk Ridge

Seasonal average of sightings (over twenty years): 68

Earliest date of sighting: August 18

Latest date of sighting: October 31

Peak migration: September and early October

Record daily count: 21 on September 29, 1997

Record seasonal count: 116 in 2010

Gyrfalcon *(Falco rusticolus)*

The largest of the falcons and most northern-ranging of all hawks is the Gyrfalcon. Few migrate far from their arctic and subarctic breeding grounds. On only eleven occasions since 1981 have they been spotted at the main overlook at Hawk Ridge; Hawk Ridge banders have trapped and banded seven since 1974. Obviously a visitor's chances of seeing one are low, but fortunately for hawk watchers, every few years a Gyrfalcon spends an entire winter in the Duluth harbor. Once birders figure out its daily patterns, it can be relatively easy to see.

As rare as it is to see a Gyrfalcon, it is even trickier to accurately identify it. This species' plumage is extraordinarily variable, ranging from nearly pure white to almost uniformly dark grayish brown. Most seen in Minnesota are gray. The size variation between sexes is extreme, with males weighing only 65 percent as much as females. A great many reports of Gyrfalcons in the harbor are misidentified Peregrine Falcons, which also winter in Duluth. Gray Gyrfalcons can be distinguished from peregrines by their lack of a bold helmet. In flight, the gyr has a longer tail and shorter, broader wings with a rounder tip. From below, the gyr's two-toned underwing can also be a useful identifying feature. Except for hawk watchers with a lot of experience identifying Peregrine Falcons of all ages from all angles, documenting a Gyrfalcon is best done with a reasonably clear photograph.

The occasional wanderings of Gyrfalcons and their reproductive success from year to year are highly dependent on availability of their favorite prey, arctic grouse called ptarmigan. Sightings of gyrs are rare in Minnesota, but the species is not rare within its remote northern habitat. Climate change is the greatest risk for the Gyrfalcon because of its narrow ecological niche, specializing on prey that is equally limited to forbidding arctic habitats.

Emperor Frederick II of Hohenstaufen, in his thirteenth-century treatise on falconry, said that the Gyrfalcon "holds pride of place over even the Peregrine in strength, speed, courage, and indifference to stormy weather." Watching any large falcon, even for a moment, is a thrilling experience. Knowing that one is a Gyrfalcon as it rockets past carries that thrill to exhilarating heights.

Gyrfalcons at Hawk Ridge

11 records since 1981

Earliest date of sighting: September 21

Latest date of sighting: November 19

Prairie Falcon *(Falco mexicanus)*

The Prairie Falcon is a rare straggler to Minnesota from arid regions of the western Plains. It has appeared at Hawk Ridge just eight times since 1979, and there are a few records of them wintering here and there in the state. Their visits to Minnesota are utterly unpredictable.

Prairie Falcons have been less intensively studied than peregrines, but some observers believe they may fly at similar or even greater speeds. They average just a bit smaller than peregrines. Paler plumage, a narrower moustachial stripe, and dark axillars (the feathers in their "wing pits") distinguish Prairie Falcons from peregrines. ◎◎

Prairie Falcons at Hawk Ridge

8 records since 1979

Earliest date of sighting: August 6

Latest date of sighting: November 10

Prairie Falcon

Visiting Hawk Ridge

As part of the Duluth city park system, Hawk Ridge is "open" from April to November, and there is never a charge to park or to visit. To see migrating hawks, visit Hawk Ridge from mid-August through October. Most raptors pass through between 10:00 a.m. and 2:00 p.m. Migration is weather dependent, with largest flights occurring when the wind is from the west or northwest. Few hawks, if any, fly in rain.

Hawk Ridge is undeveloped habitat without permanent structures. The main overlook at Hawk Ridge is on a steep, exposed bluff. On hot days, this area can be warmer than the surrounding city; on cold days, the wind can be extreme. Dress for the weather, bring extra clothes, and apply sunblock. If you stay at the main overlook, most comfortable shoes are appropriate, but to walk the trails (which are fairly rugged) wear sturdy sneakers or hiking shoes. Restrooms are not available, except during the middle of September, so using a rest-room before arriving will make your visit more comfortable. If you don't wish to stand throughout your visit, boulders are available on a first-come, first-served basis. Many people bring chairs.

Binoculars are the most important optics for finding and identifying hawks and for counting groups of them. You will be able to see more hawks in view at the same time with 7× or 8× binoculars, but you will see them closer with 10×. If you don't have binoculars, loaners for use during your visit are available free from Hawk Ridge staff from August 15 through October 31. A spotting scope can help with challenging identifications, but it is usually much less effective than binoculars because the narrow field of view makes finding and counting difficult.

From September 1 through October 31, a migration interpreter is often walking about, pointing out birds and helping with identification. It may help to study the basics of hawk identification in advance. Flight identification sheets and brochures are available on the Hawk Ridge website (www.hawkridge.org). The most useful field guides to bring are Jerry Liguori's *Hawks from Every Angle* and *Hawks at a Distance;* Clay Sutton, Pete Dunne, and David Sibley's *Hawks in Flight;* and Brian Wheeler and William Clark's *A Photographic Guide to North American Raptors.* If you want assistance with hawk identification, station yourself at the main overlook to hear staff and visitors call out birds and identify them. If you want a quiet, more solitary hawk-watching experience, Hawk Ridge's extensive trail system will lead you to more secluded spots.

When the banding station sends hawks to the main overlook for release, volunteers will call attention to them. Regularly scheduled programs are also presented, some at night, to learn about owl migration and, when possible, to show banded owls before release. You can learn about these during your visit, or check ahead of time at www.hawkridge.org.

Because Hawk Ridge is part of Duluth's Parks and Recreation Department, it is left as natural as possible. The access road is closed in winter. During spring and summer, the birds near the main overlook are typical species of second-growth forest and open rocky habitat. Many people enjoy the pleasant hike through the diverse habitats between Hawk Ridge and Seven Bridges Road. That hike includes a long hill, from Hawk Ridge at the peak descending to the end of Seven Bridges Road, where it meets Lester Park.

For more information about Hawk Ridge, visit www.hawkridge.org.

Acknowledgments

I would like to thank the many people who did so much to establish Hawk Ridge and to ensure that it will be protected in perpetuity, beginning with Jack Hofslund, who first brought our hawk migration to state and national attention.

I am grateful to Karl Bardon, David Carman, Alison Clarke, Kim Eckert, David Evans, Molly Evans, Jan and John Green, David Grosshuesch, Frank Nicoletti, Jerry Niemi, Julie O'Connor, Henry Roberts, Wayne Russ, Sparky Stensaas, Koni Sundquist, and Debbie Waters, who for years, each in their own important ways, have helped make Hawk Ridge what it is.

Since I moved to Duluth in 1981, one person has been my biggest source of inspiration in my advocacy for birds. Janet Green has quietly become the Rachel Carson of Minnesota. Doggedly working behind the scenes with governmental agencies, nonprofits, corporations, and individuals, Jan's in-depth understanding of both birds and the bureaucratic mazes constructed by conservationists and their opponents ever fills me with awe. Her meticulous scholarship, fair-mindedness, tireless work ethic, and unfailing patience and courtesy earn the respect of all who know her. Every time I have needed help dealing with a bird issue, Jan has been there for me, and this book is dedicated to her.

Laura Erickson has been writing and speaking about birds and promoting their conservation for thirty-five years. Her daily radio spot and iTunes podcast *For the Birds* airs nationally. She has written four previous books, including *Twelve Owls* (Minnesota, 2011) and *Sharing the Wonder of Birds with Kids,* which received the National Outdoor Book Award. She lives in Duluth.

Betsy Bowen is a renowned woodcut printmaker and painter. She has illustrated several books, among them *Great Wolf and the Good Woodsman, Borealis, Wild Neighborhood, Big Belching Bog,* and *Twelve Owls,* all published by the University of Minnesota Press. She lives in Grand Marais, Minnesota.

Also Published by the University of Minnesota Press

Twelve Owls
Laura Erickson
Illustrations by Betsy Bowen

For the Birds: An Uncommon Guide
Laura Erickson
Illustrations by Jeff Sonstegard

Sharing the Wonder of Birds with Kids
Laura Erickson
Illustrations by Kathryn Marsaa

Big Belching Bog
Phyllis Root
Illustrations by Betsy Bowen